SPIRIT OF

NORFOLK WINDMILLS

CHRIS HERRING

PiXZ

First published in Great Britain in 2010

British Library Cataloguing-in-Publication Data
A CIP record for this title is available from the British Library

ISBN 978 0 85710 026 9

PiXZ Books
Halsgrove House, Ryelands Industrial Estate,
Bagley Road, Wellington, Somerset TA21 9PZ
Tel: 01823 653777
Fax: 01823 216796
email: sales@halsgrove.com

An imprint of Halstar Ltd, part of the Halsgrove group
of companies Information on all Halsgrove titles is
available at: www.halsgrove.com

Printed and bound in China by Toppan Leefung Printing Ltd

Introduction

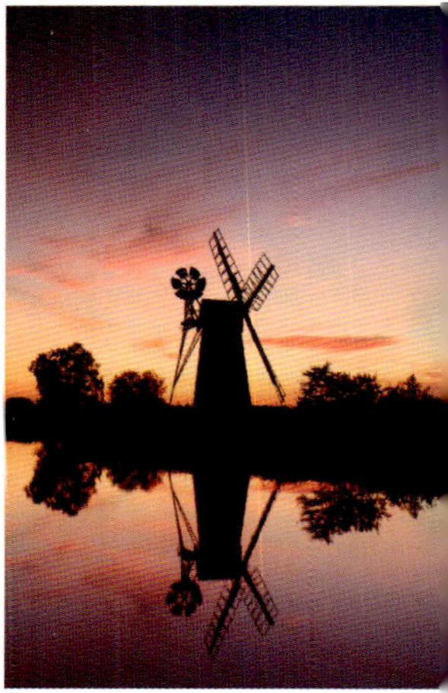

The iconic mills of Norfolk are scattered all across the county. In the nineteenth century at their peak there were around 700 working windmills in Norfolk and those that survive are responsible for how the Norfolk landscape looks today.

Those best known in the county are the windmills of the Norfolk Broads. The engineering of dykes and associated windpumps helped to drain the marshland and allow farming and settlements to grow. This process began over 250 years ago, and whilst over the years the drainage mills of the Broads have gradually been replaced with more efficient diesel and electric pumps, plenty of examples of the original mills can still be found in both working and derelict states.

Whether the mills were originally built to grind corn or to drain water the basic process of harnessing wind power through the use of sails is still the same, and this book showcases a selection of my personal favourite windmills and drainage mills that can be found in Norfolk.

Chris Herring

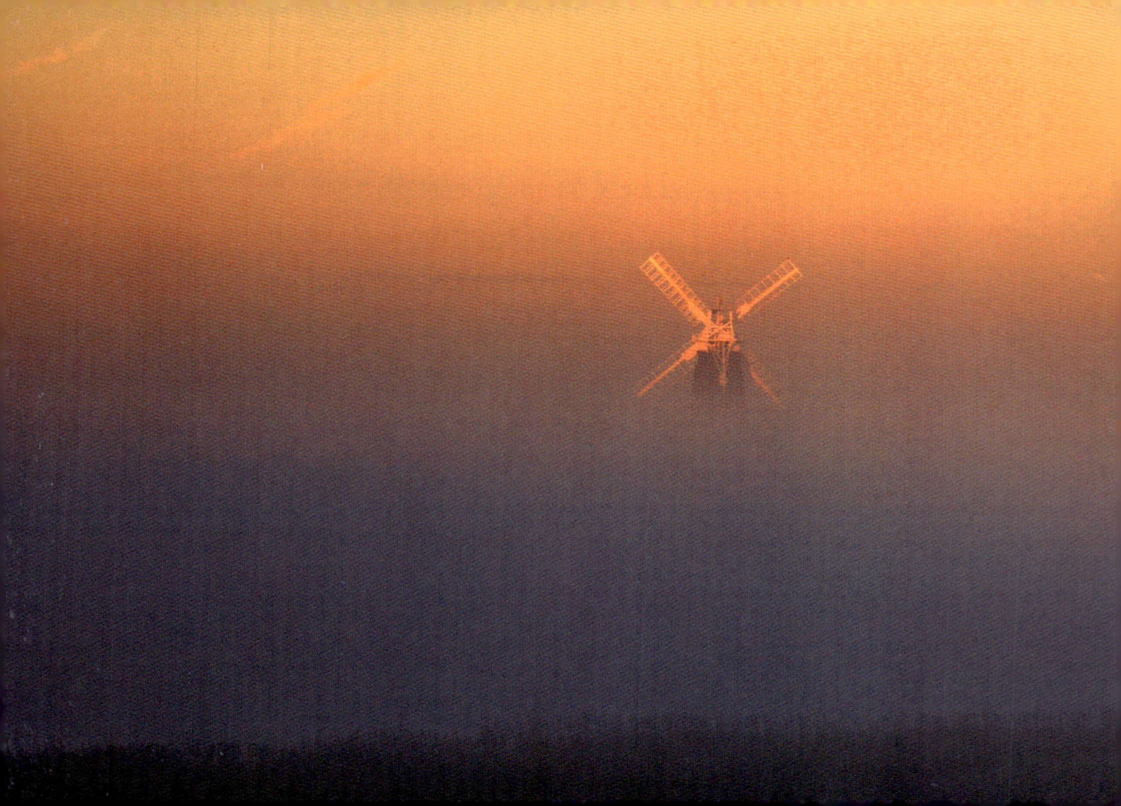

The newly restored Berney Arms drainage mill rising above the morning mist over Halvergate Marshes at dawn on the Norfolk Broads.

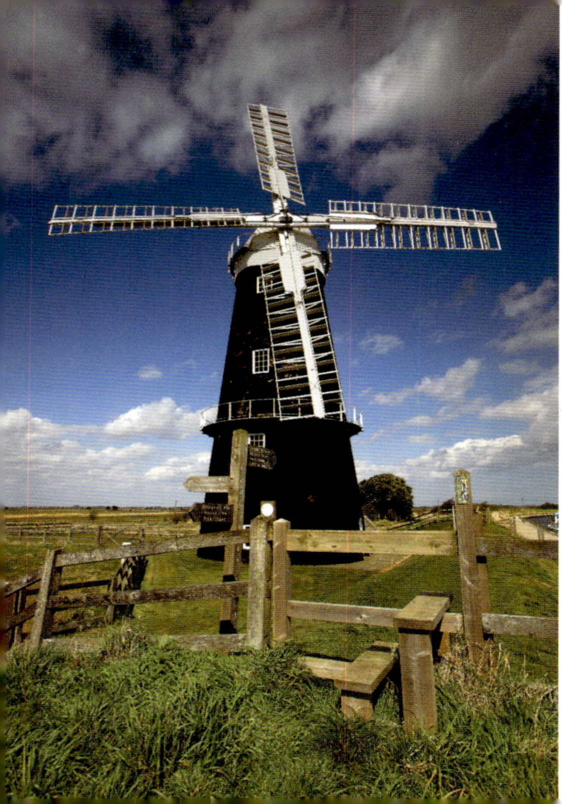

Berney Arms Mill stands in a remote location on the Halvergate Marshes. Access to the mill is via a two-hour walk across the marshes, a request train stop, or by boat.

Opposite page:
Berney Arms Mill sits alongside the River Yare and is the tallest mill on the Norfolk Broads. Following a lengthy restoration period the sails were finally returned to the mill in the summer of 2007.

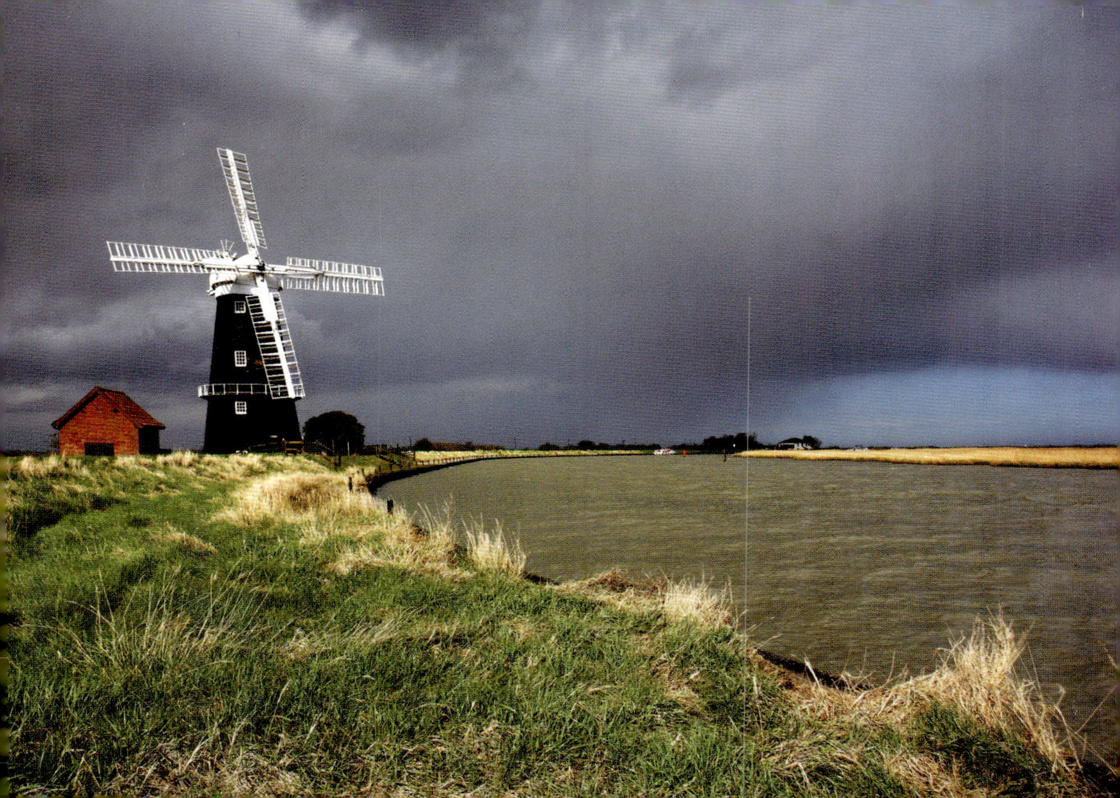

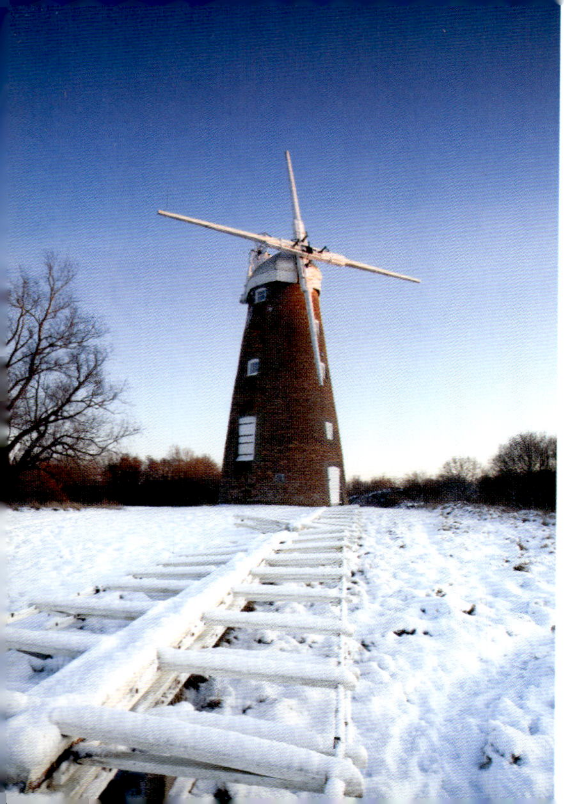

Billingford Tower Mill, close to the Suffolk border, was photographed during snowfall and whilst restoration work was being undertaken on the main sails.

Opposite page:
Boardman's open-framed timber drainage mill on the banks of the River Ant, captured here on a stormy evening on the Norfolk Broads.

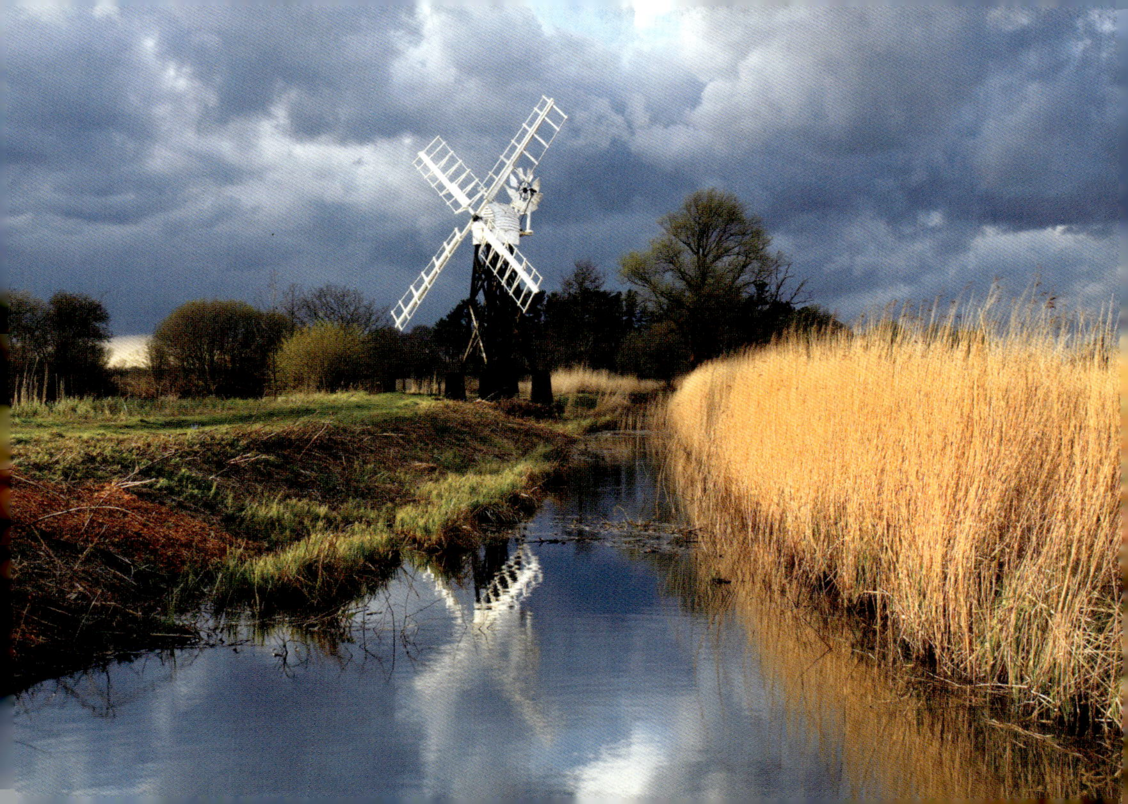

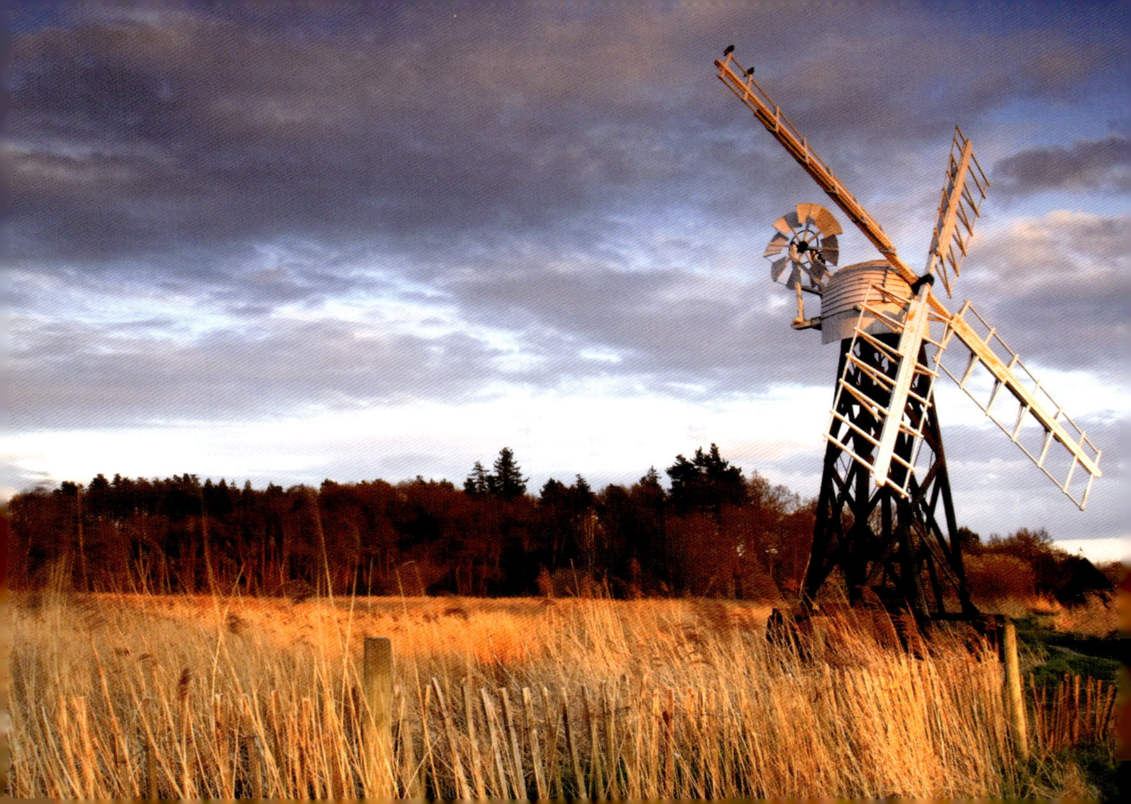

Open-framed timber drainage mills, or skeleton mills as they were often known, were a less expensive alternative to the traditional brick-built mills. Here Boardman's Mill is reflecting in the River Ant on a bright sunny day.

Opposite page:
Late evening light illuminates the reeds beside Boardman's drainage mill.

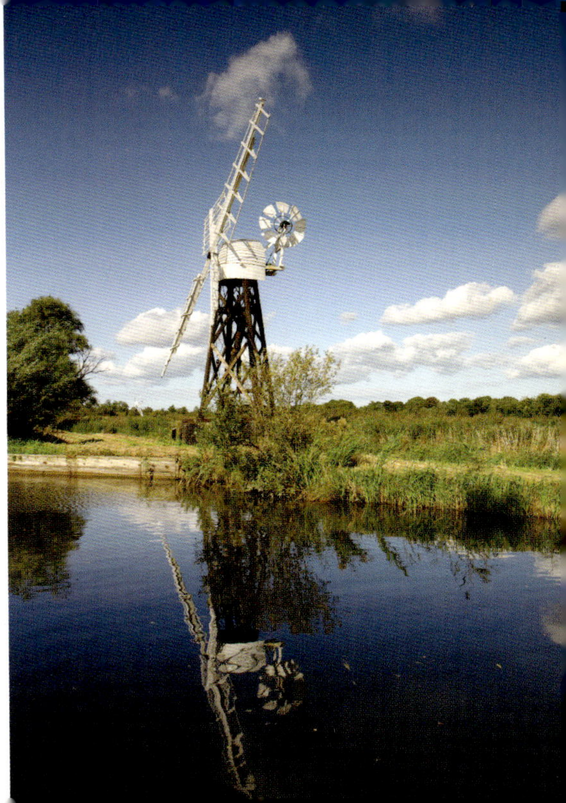

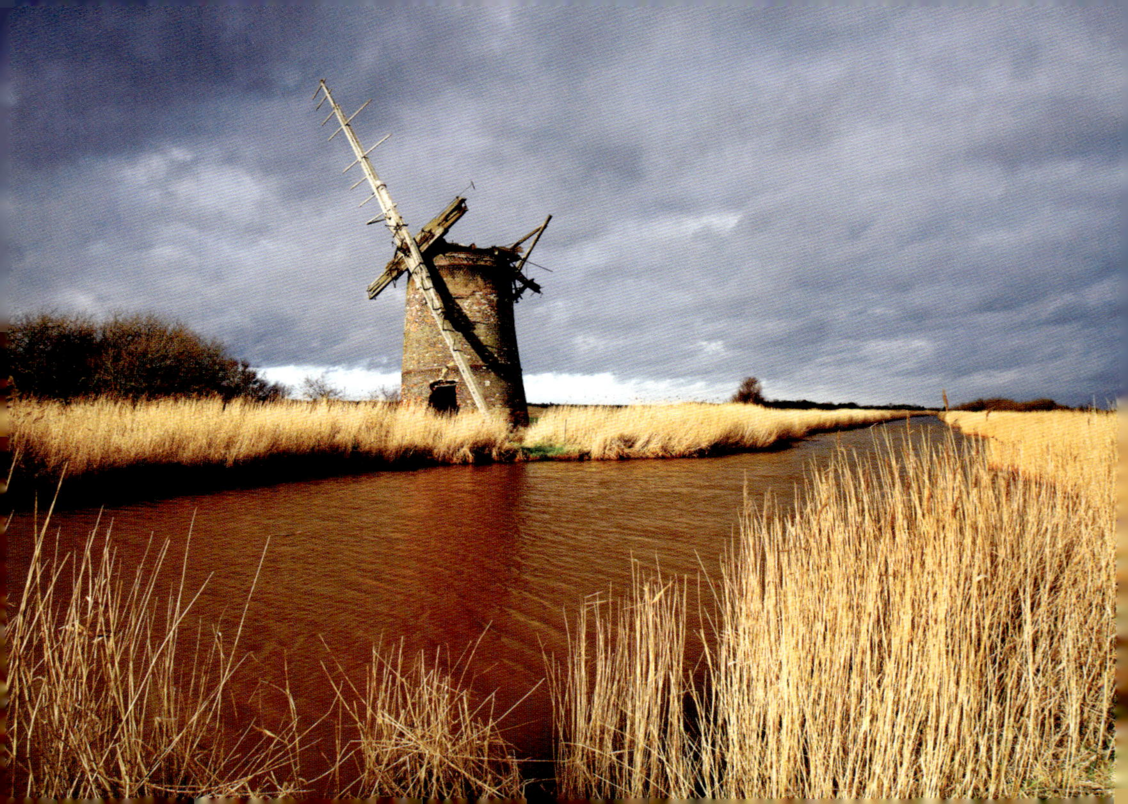

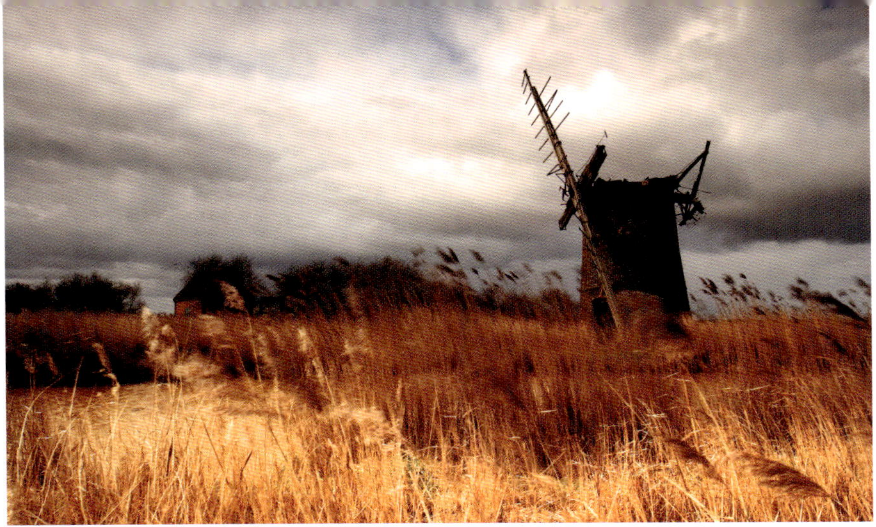

The derelict Brograve drainage mill was originally built in 1771. It is believed to have stopped working sometime in the 1930s. To the left can be seen the modern pumping station that now takes the mill's place.

Opposite page: The derelict drainage mill of Brograve on the Waxham New Cut always photographs well against a moody sky. What I like most about this location is the wonderful brown colour of the peat-stained water.

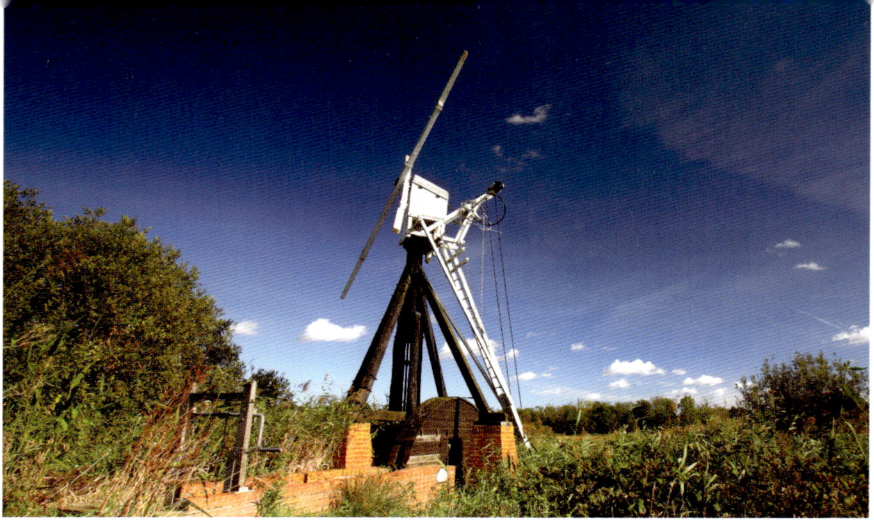

Clayrack drainage mill is one of only two 'hollow post' mills left on the Broads. Now standing beside the River Ant at How Hill, the mill originally stood on Ranworth Marshes before it was moved to its current position in 1981 to undergo restoration work.

Opposite page: Burnham Overy tower mill, built in 1816, originally worked in conjunction with the nearby Burnham Overy lower watermill. Today the windmill is used for holiday accommodation.

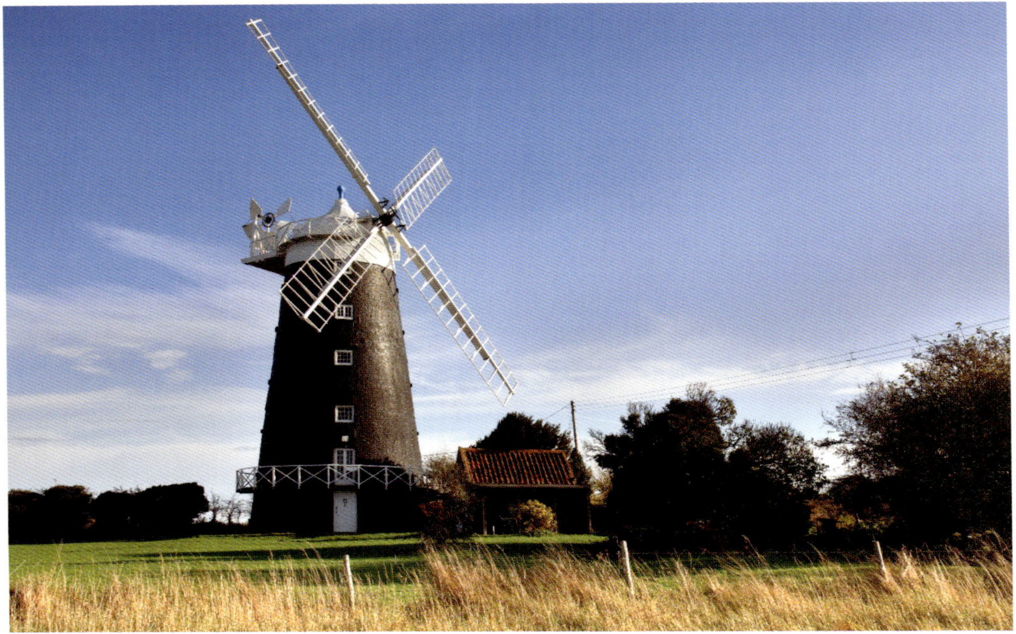

Cley tower mill and the village of Cley-next-the-Sea photographed on a bright sunny day with the whispering reeds blowing in the wind.

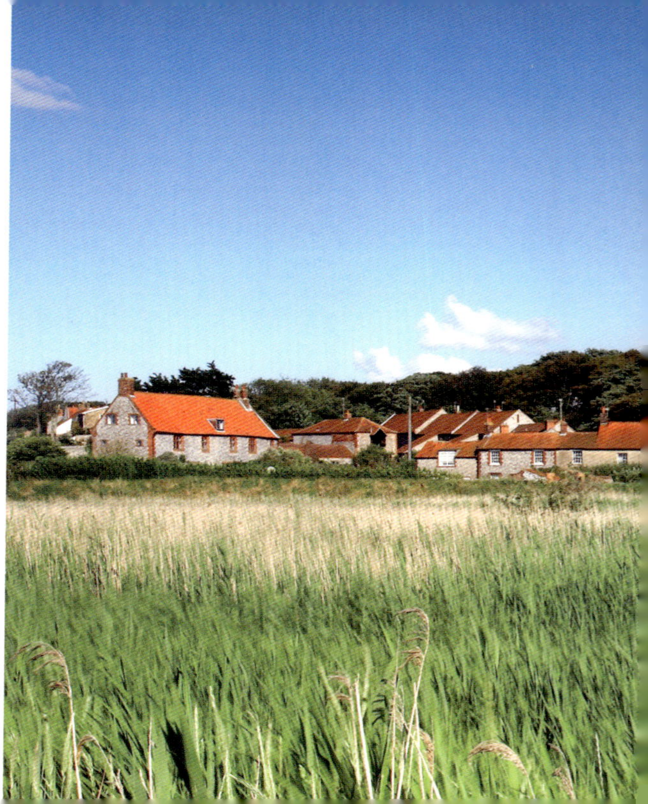

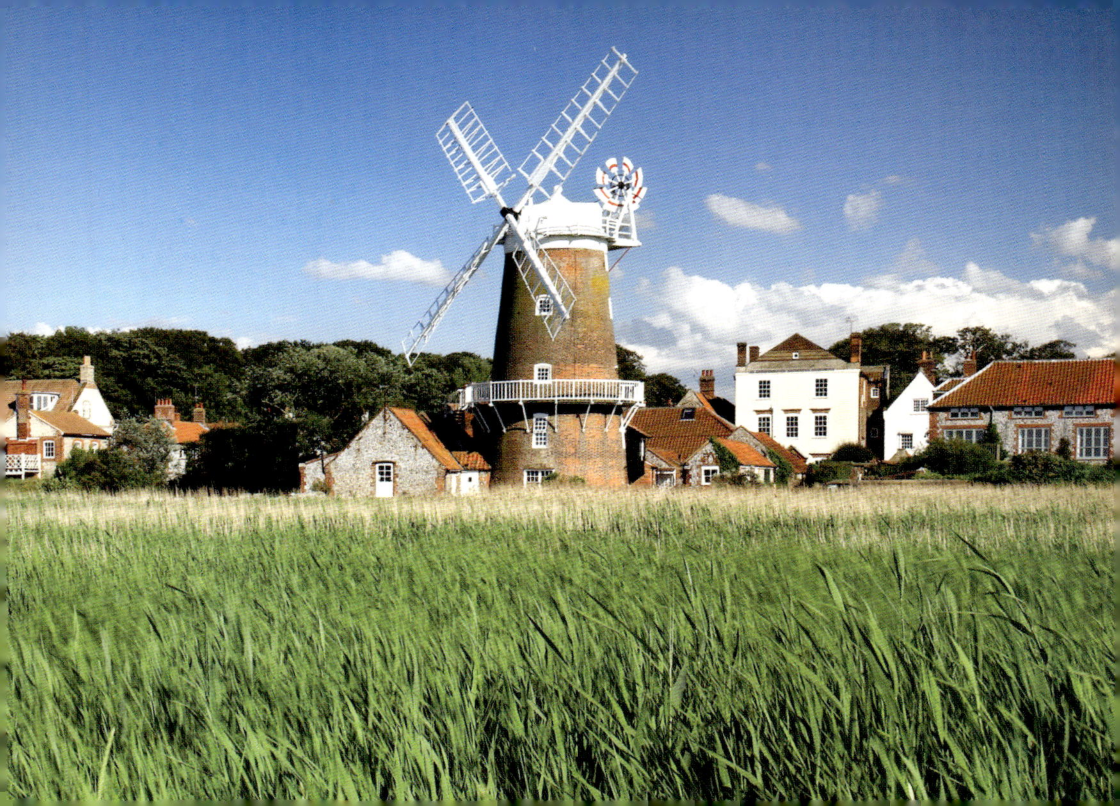

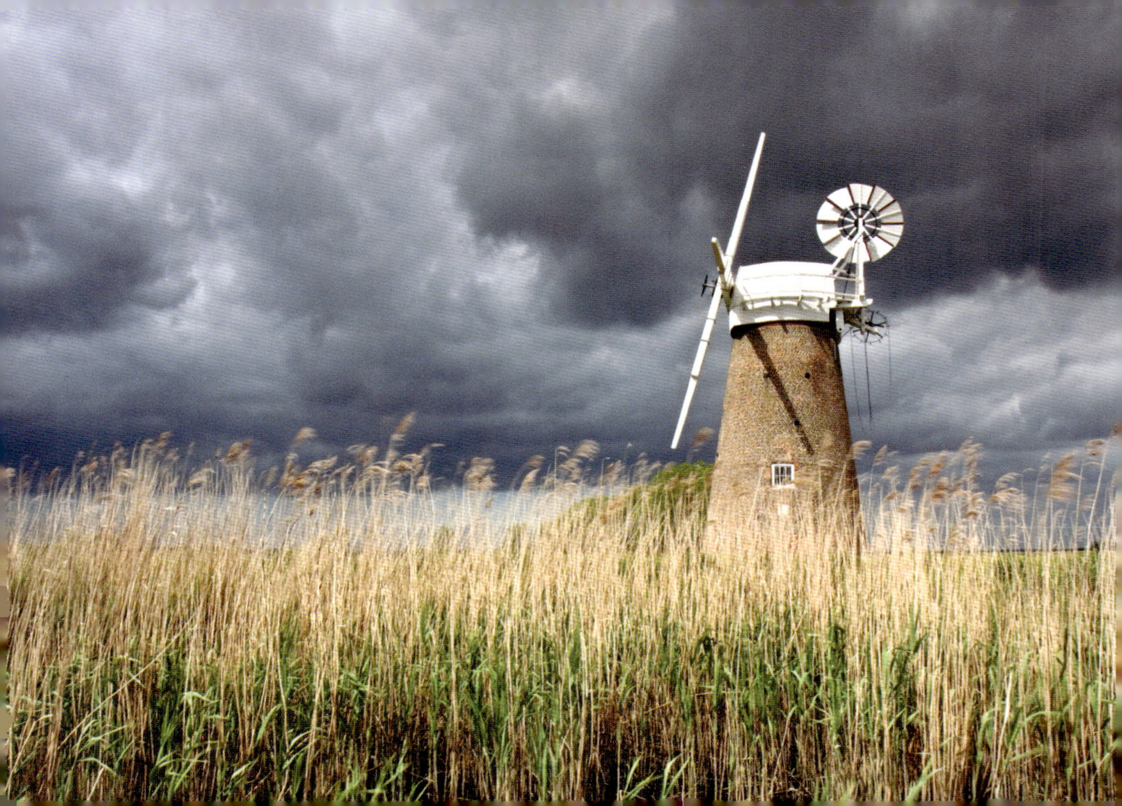

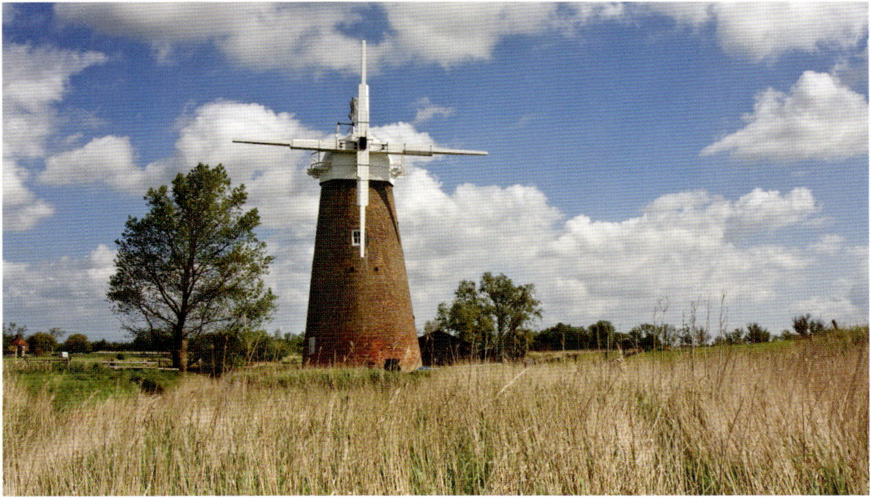

Hardley Mill on a bright spring day on the Norfolk Broads.

Opposite page: Hardley drainage mill was originally built in 1874. In 1991 restoration work began and in 2009 the sails and cap were finally returned to the mill. Here the stocks from the main sails can be seen against a stormy sky.

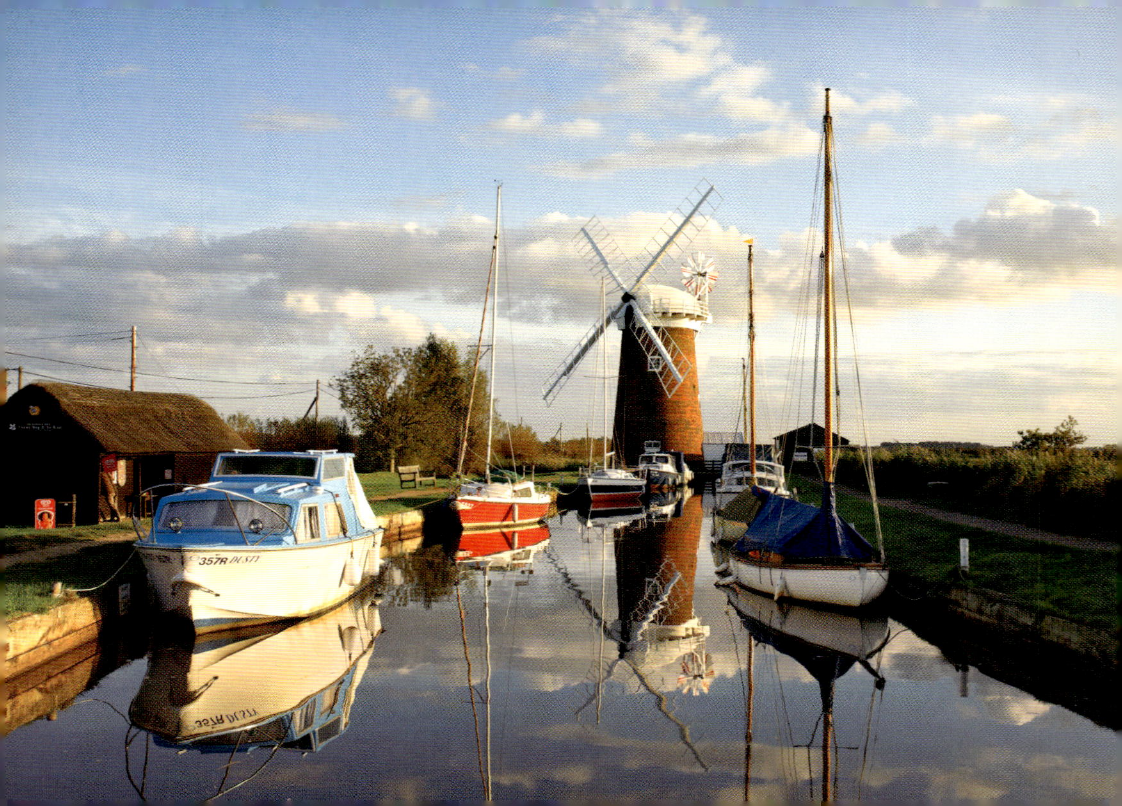
357R DUSTY

Winter reeds being blown by the wind in front of Horsey drainage mill.

Opposite page:
Horsey drainage mill reflecting in the calm still waters of Horsey Staithe on the Norfolk Broads.

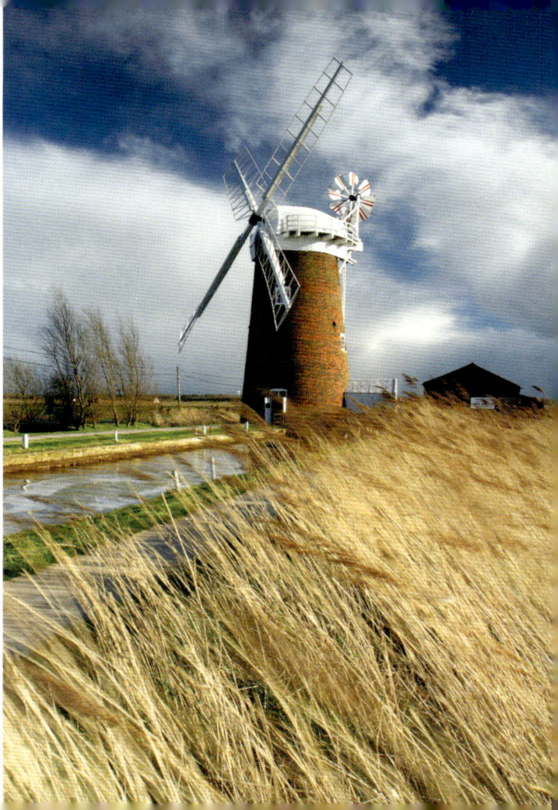

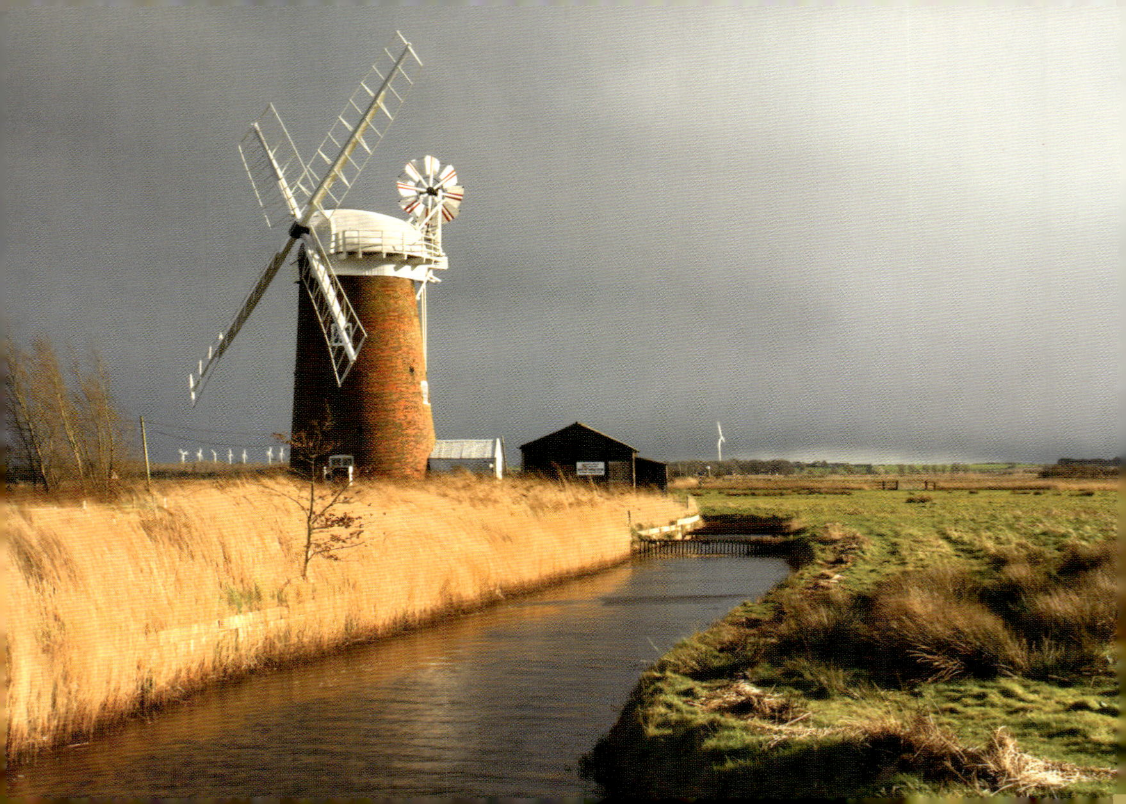

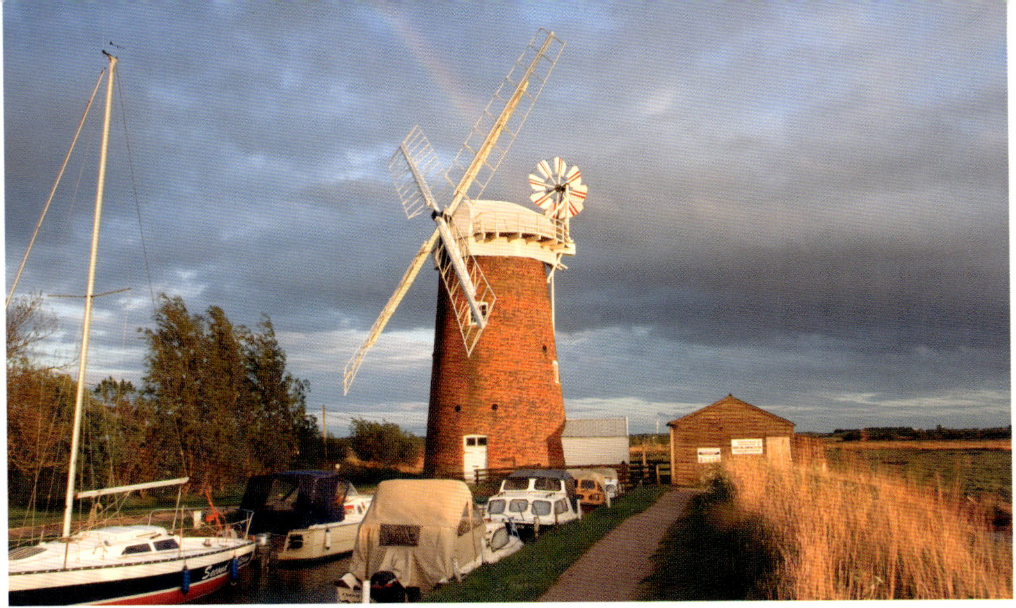

A rainbow appears above Horsey Mill during a passing storm.

Opposite page: Horsey Mill in the foreground with the modern wind farm at Winterton in the far distance.

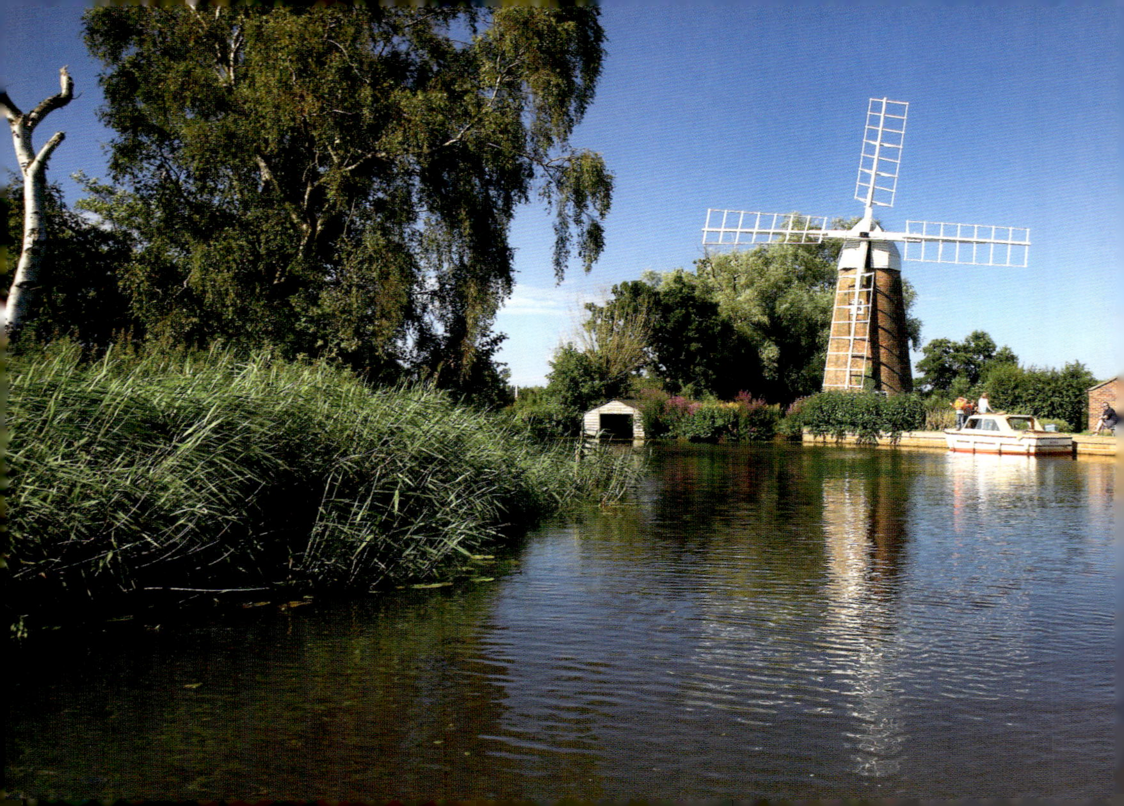

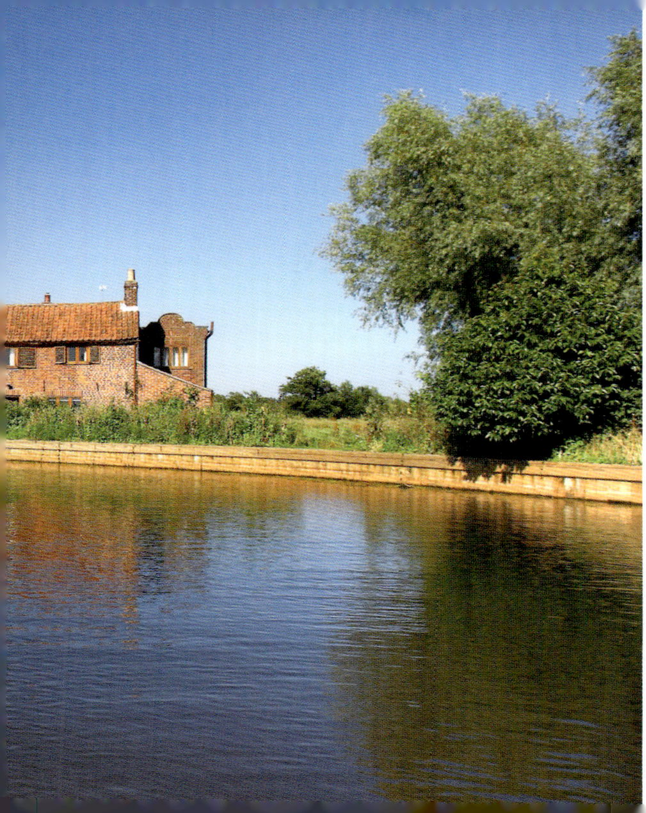

The picturesque Hunsett drainage mill captured here from a boat on the River Ant.

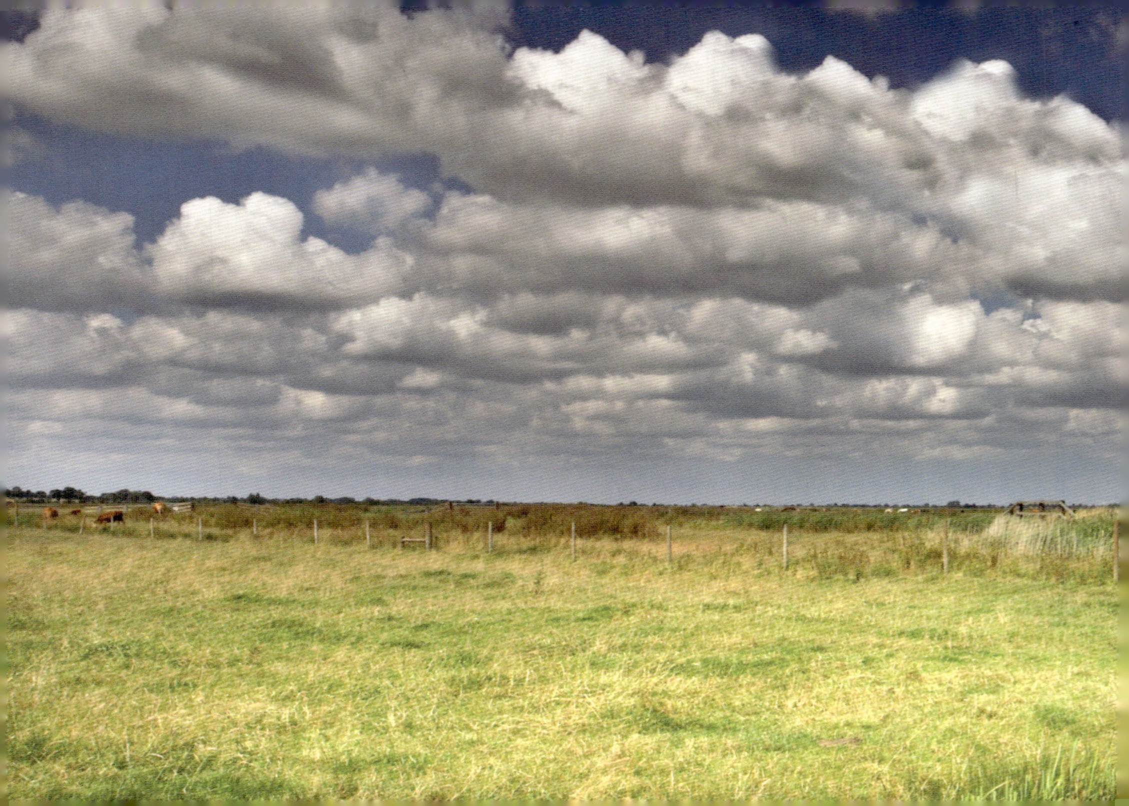

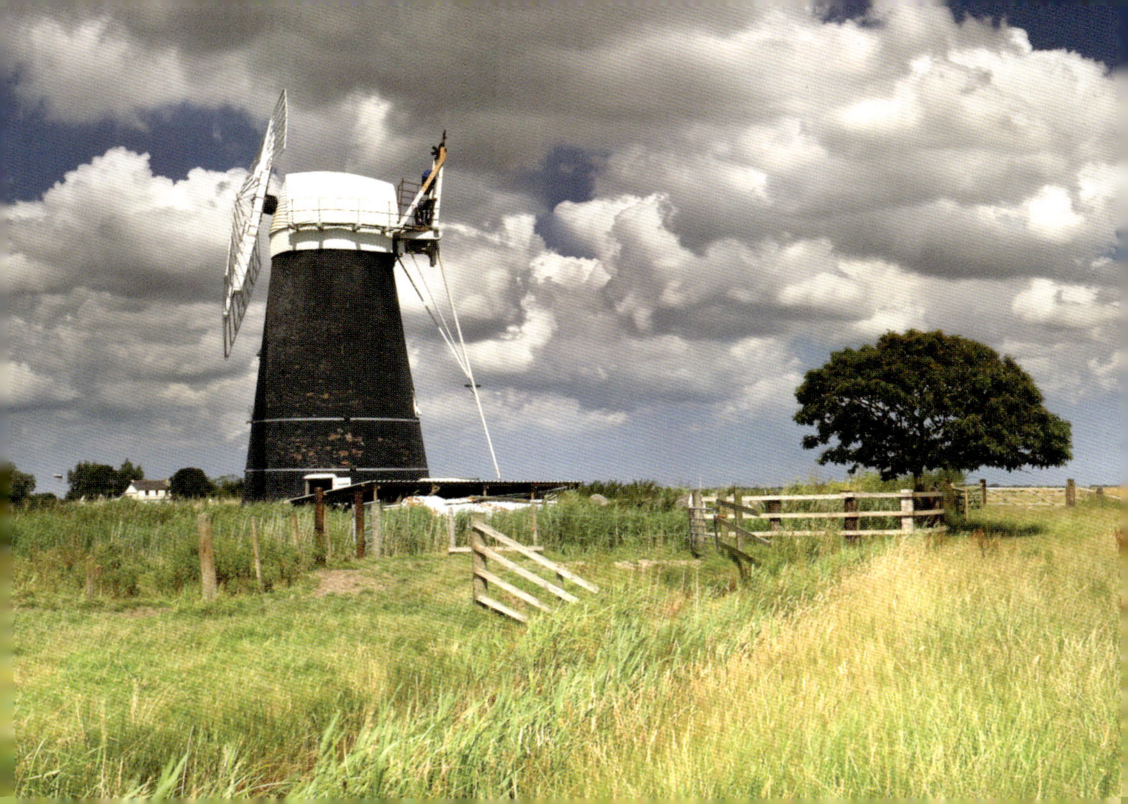

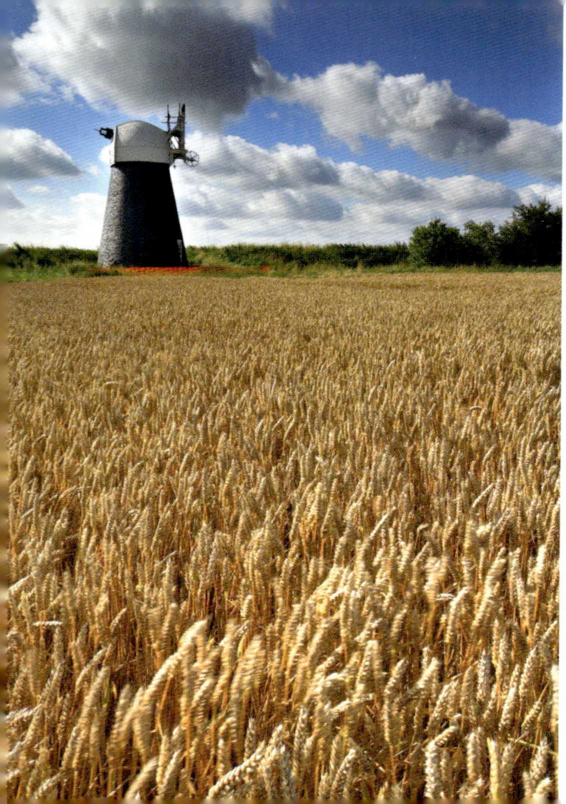

Previous page:
Mutton's drainage mill, located at Fleet Dyke, being worked on by millwrights preparing to fit a new tail fan.

Runham Swim drainage mill sits among wheatfields on a summer's day on the Norfolk Broads.

Opposite page:
The partially restored black-painted Runham Swim drainage mill on a stormy morning next to the River Bure.

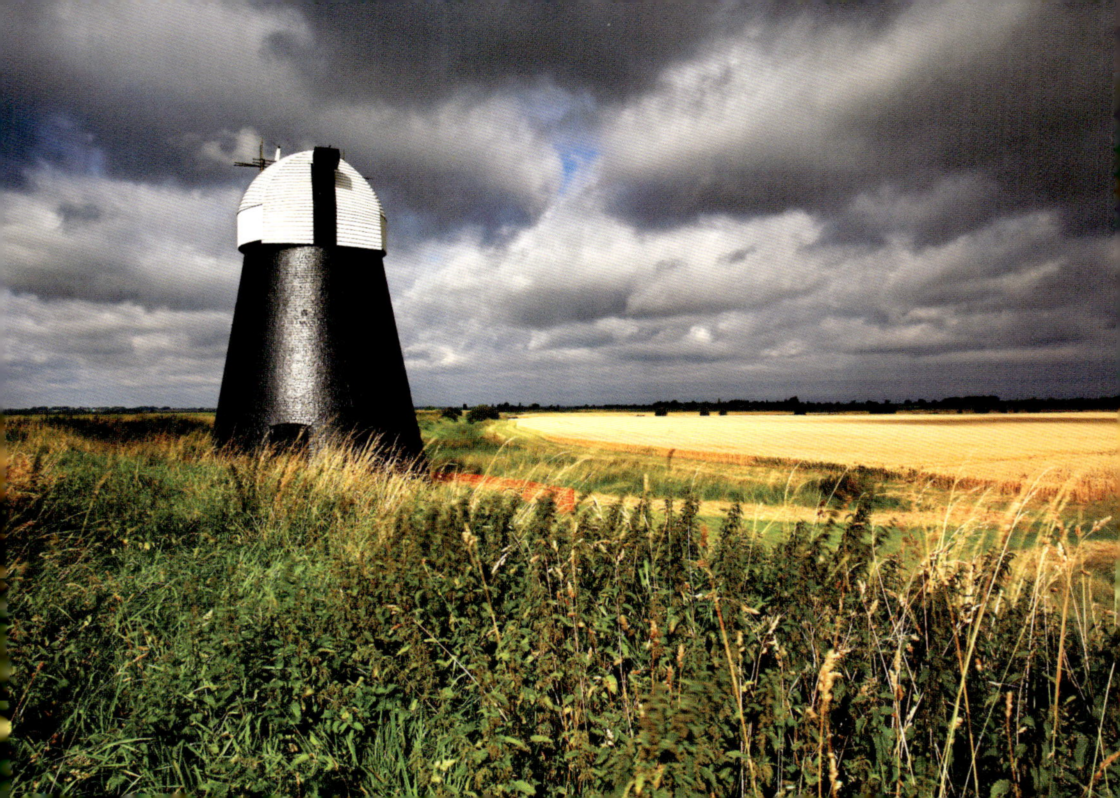

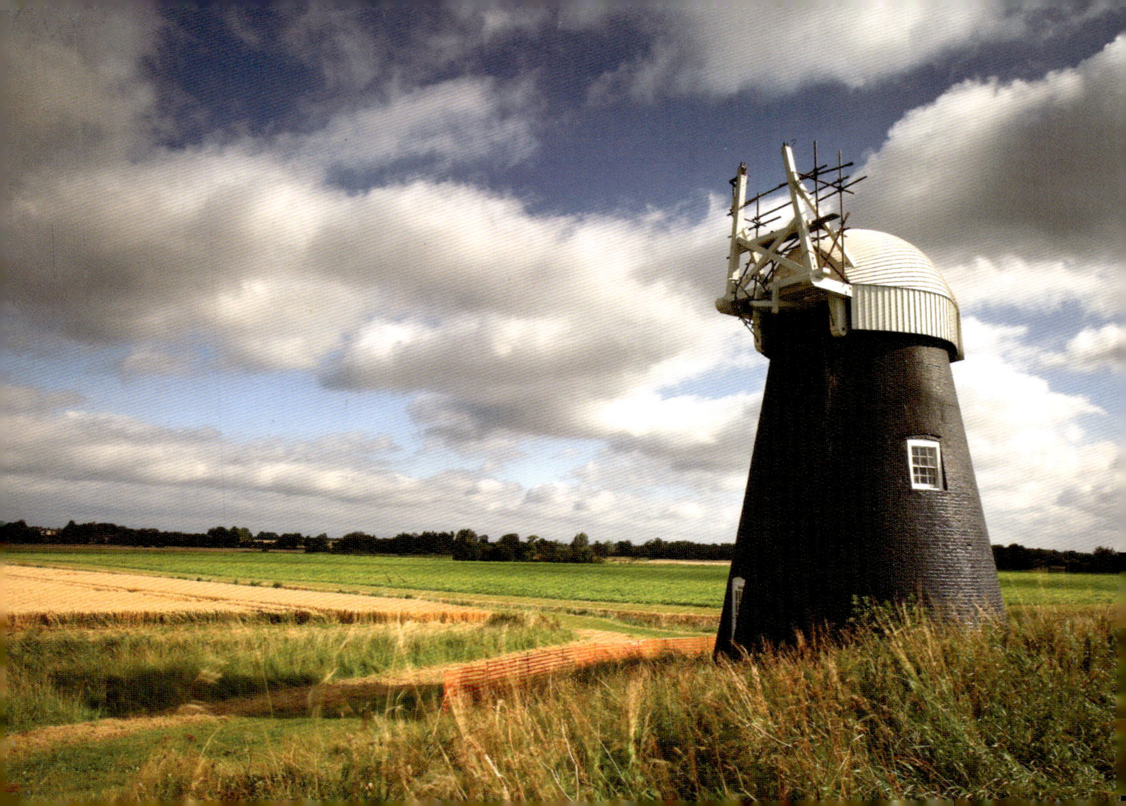

Runham Swim Mill, also known as Child's Mill is in the hands of the Norfolk Windmills Trust. The mill has been partially restored and further funding is currently being sought for complete restoration.

Opposite page:
St Benet's Level drainage mill on a bright spring afternoon.

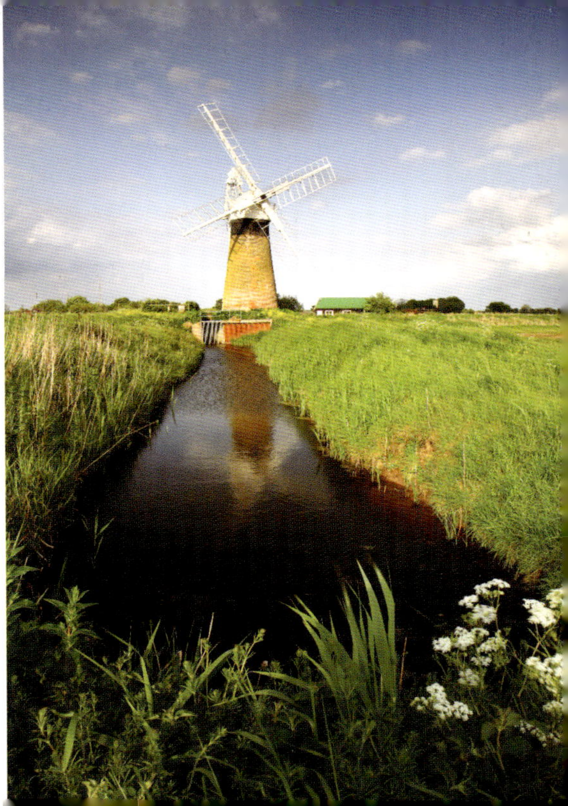

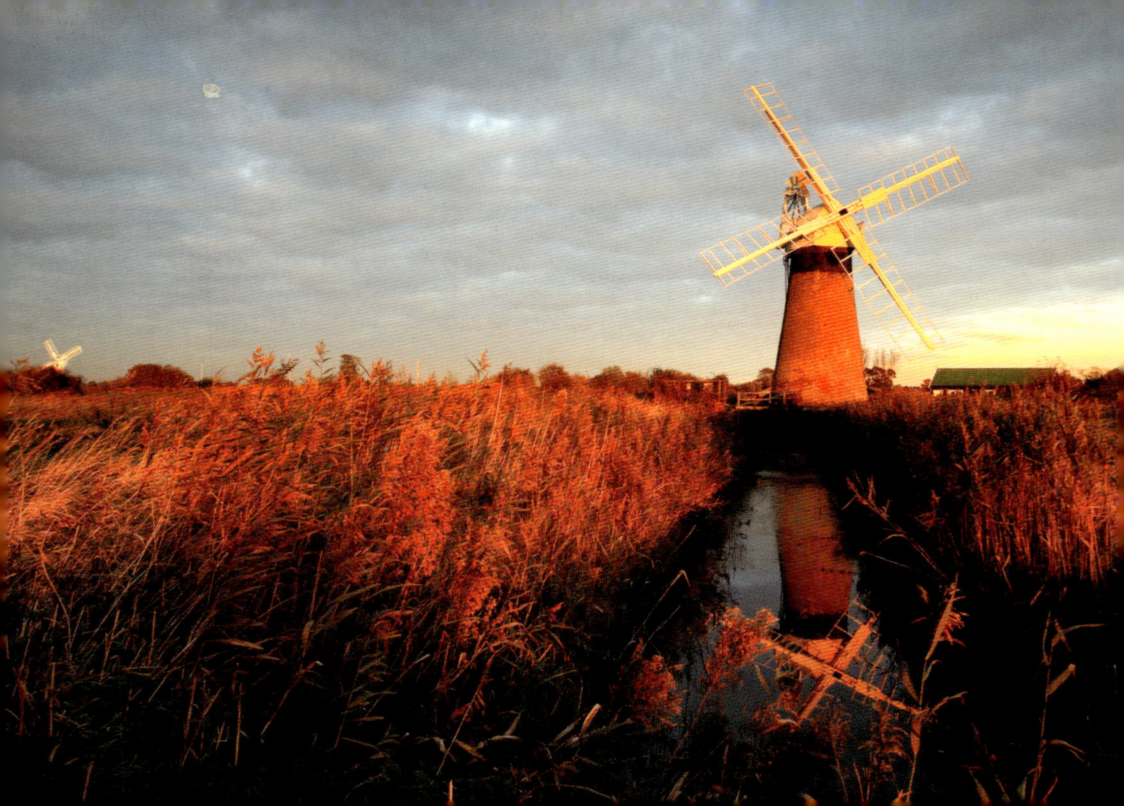

St Benet's Level drainage mill sits on the banks of the River Thurne and is owned by Crown Estates. The mill was originally built around 1775 to drain the surrounding marshland.

Opposite page:
Late afternoon light illuminates St Benet's drainage mill with the sails of Thurne Mill in the distance to the left.

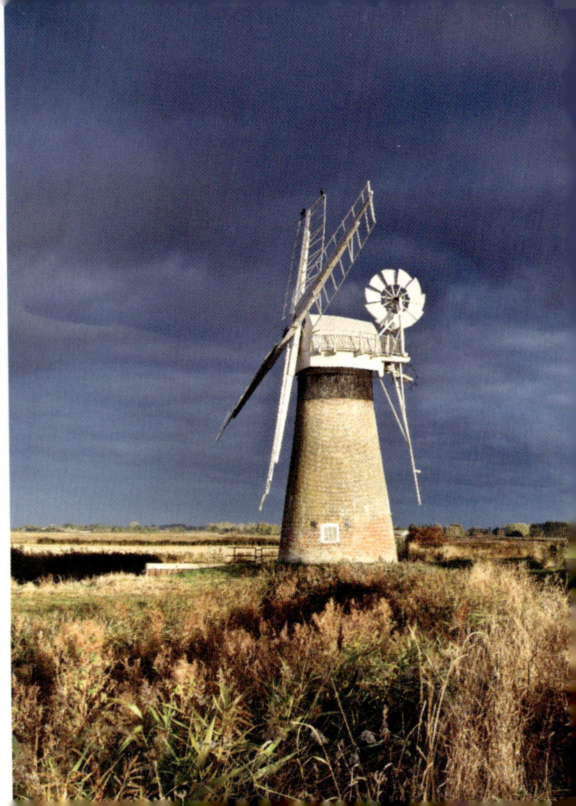

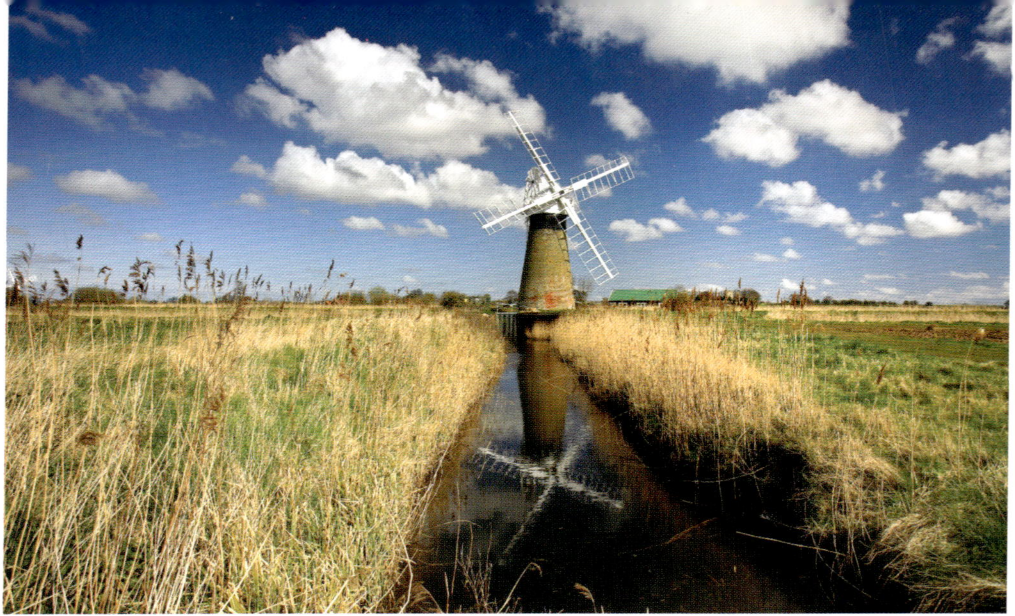

A spring afternoon besides the River Thurne on the Norfolk Broads
with St Benet's Level drainage mill reflecting in a nearby dyke.

Opposite page: Powerful storm clouds gather over St Benet's Mill shortly after dawn.

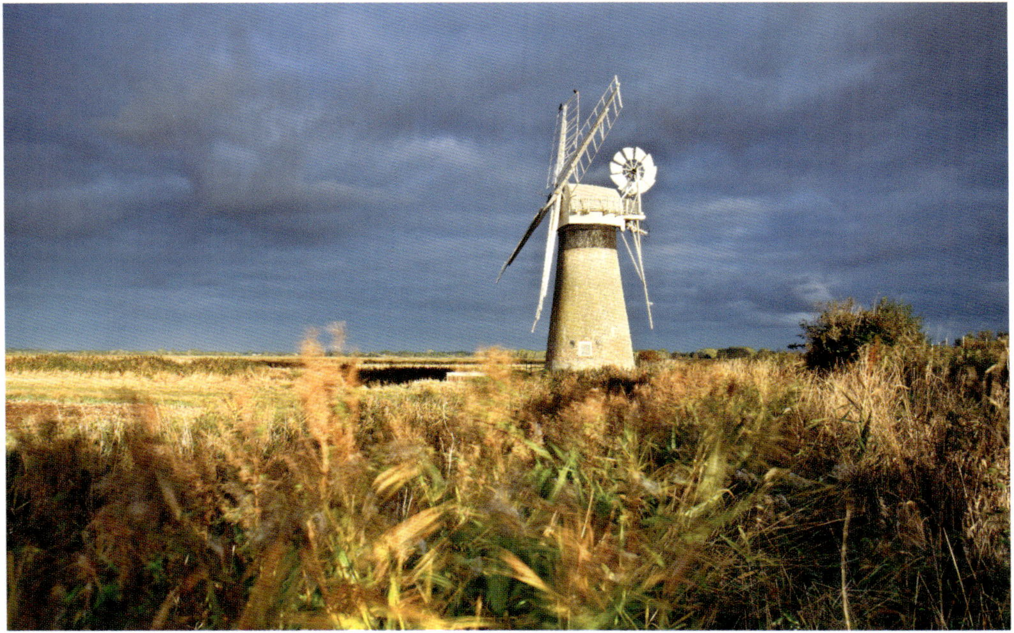

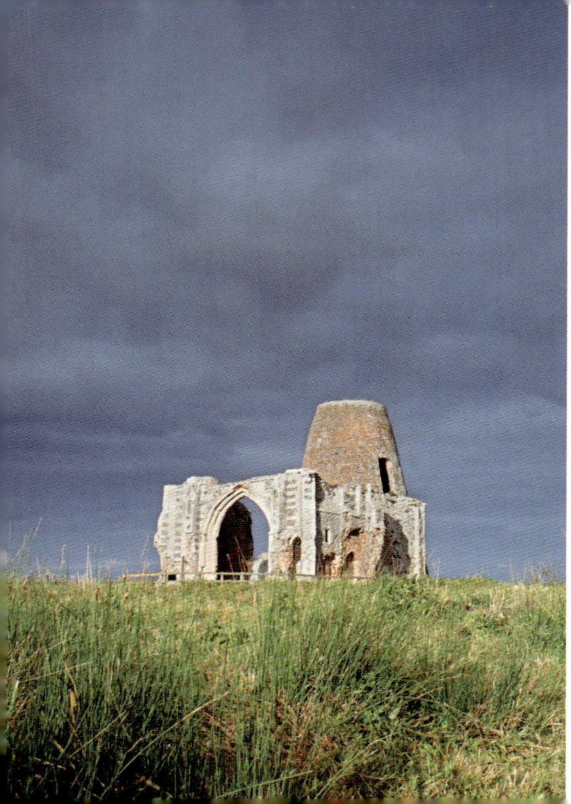

St Benet's Abbey Mill is thought to have been built between 1728–1735. What makes this mill unique is that it was built into the remains of the gatehouse of the former abbey that stood on this site.

Opposite page:
A dramatic winter hoarfrost covers the surrounding marshes at St Benet's Abbey in tiny delicate ice crystals.

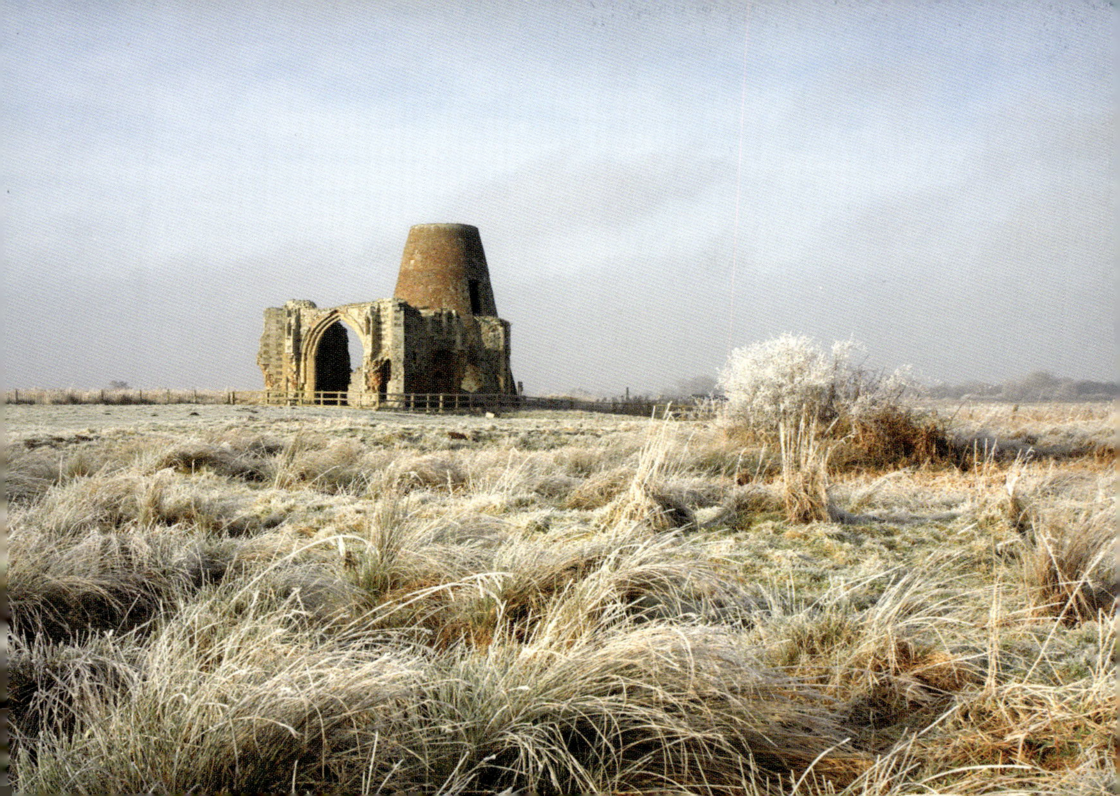

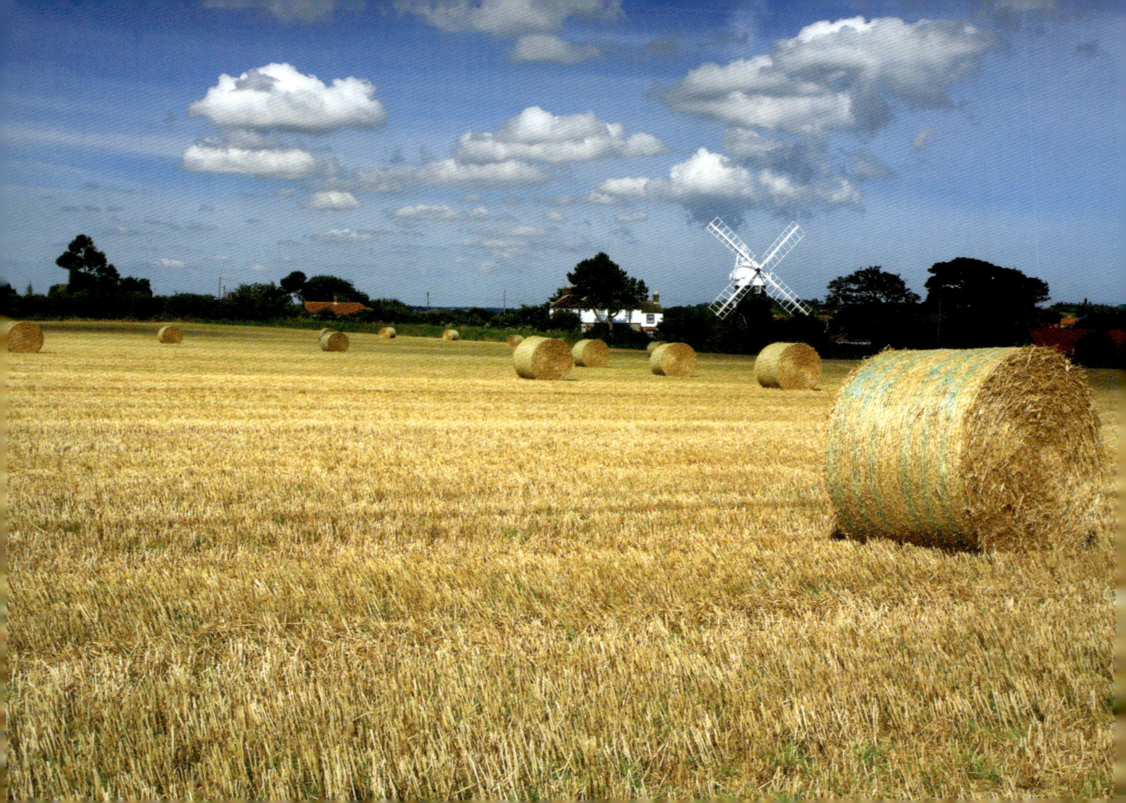

The motion of winter hoarfrosted reeds is captured here in front of Thurne drainage mill on the Norfolk Broads.

Opposite page:
Stow Mill lies beyond a field of straw bales on a summer's afternoon in the Norfolk countryside. The mill was built in 1827 as a flour mill; today it is open to the public as a tourist attraction.

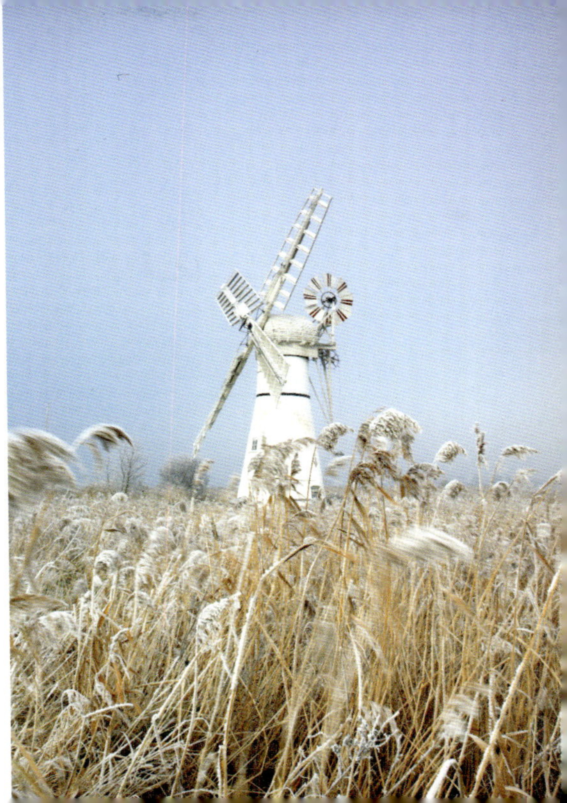

This was one of the best sunrises I have ever witnessed. For around ten minutes nature put on an amazing performance of colours over the River Thurne, the sky turned from bright oranges and reds to pinks and purples before quickly fading, with rain setting in for the rest of the day.

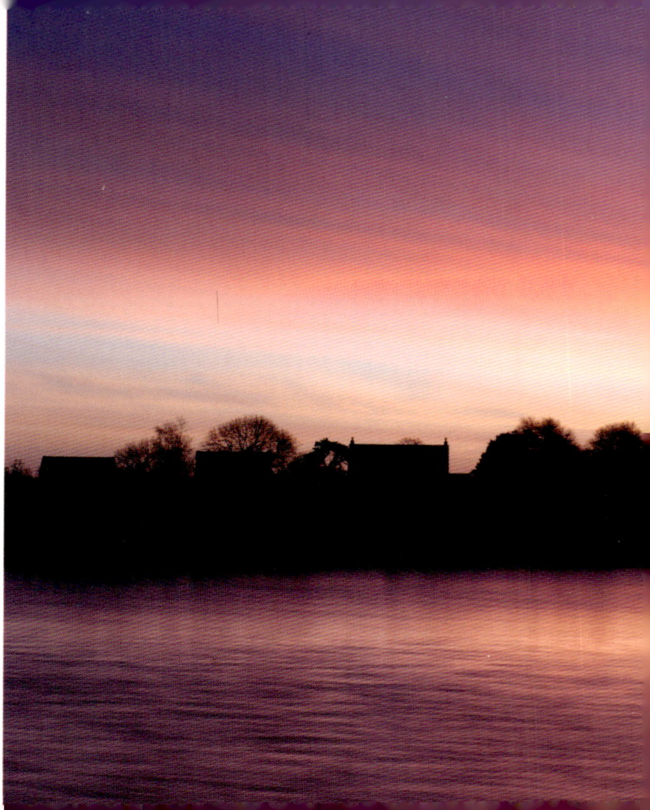

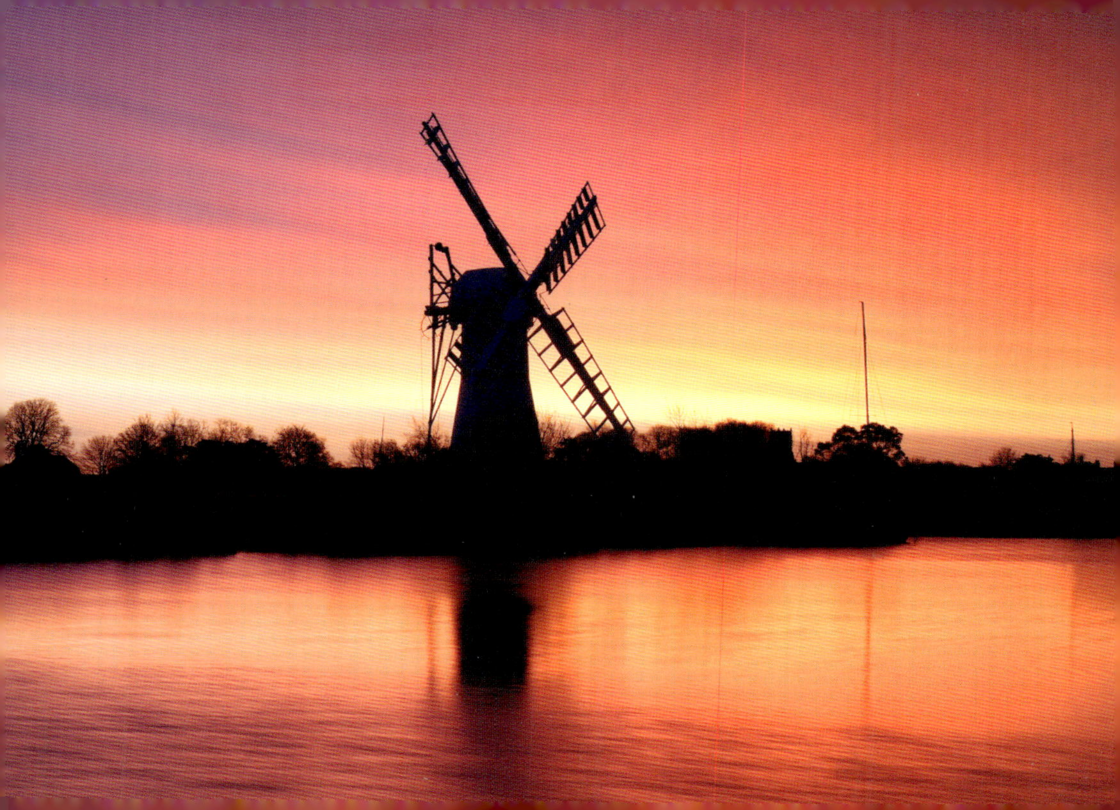

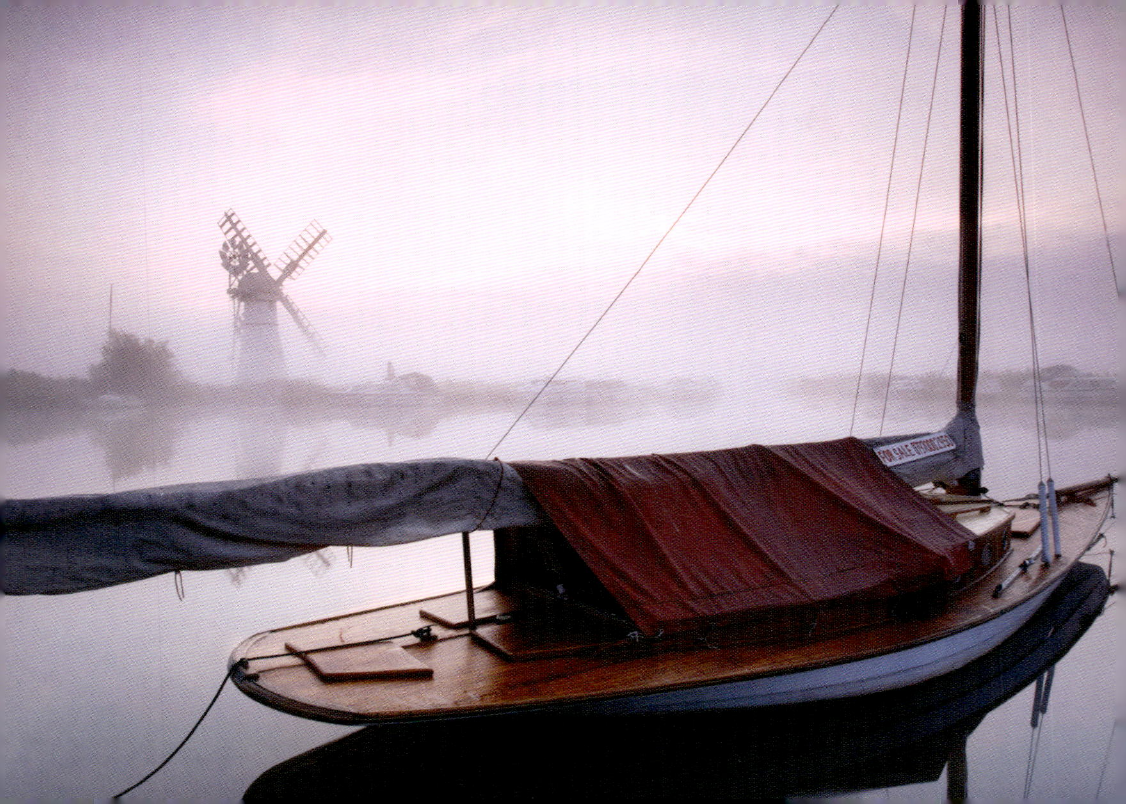
FOR SALE

The white-painted windpump of Thurne Dyke is one of the most distinctive on the Norfolk Broads. The mill was built in 1820 and after lying derelict for a number of years it was eventually saved from destruction by Bob Morse in 1949 and is now in the hands of the Norfolk Windmills Trust.

Opposite page:
A calm mystical morning captured on the River Thurne with a wooden sailing boat in the foreground and Thurne Dyke windpump on the far side of the river.

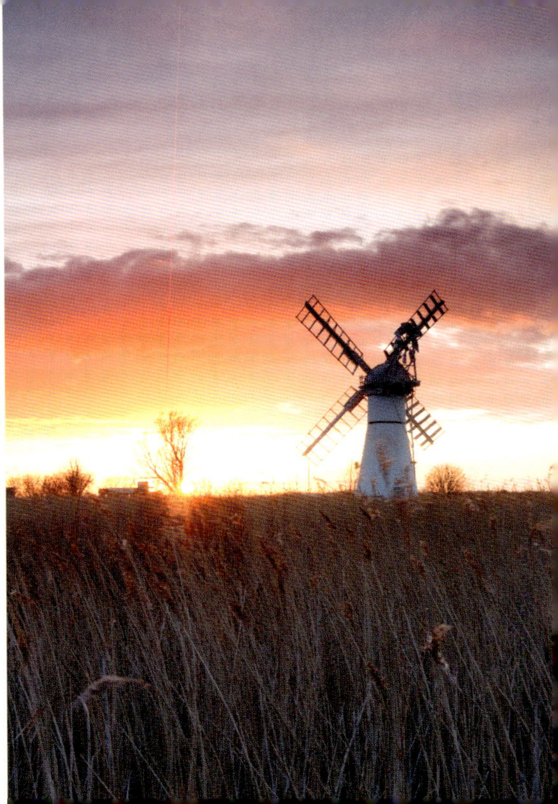

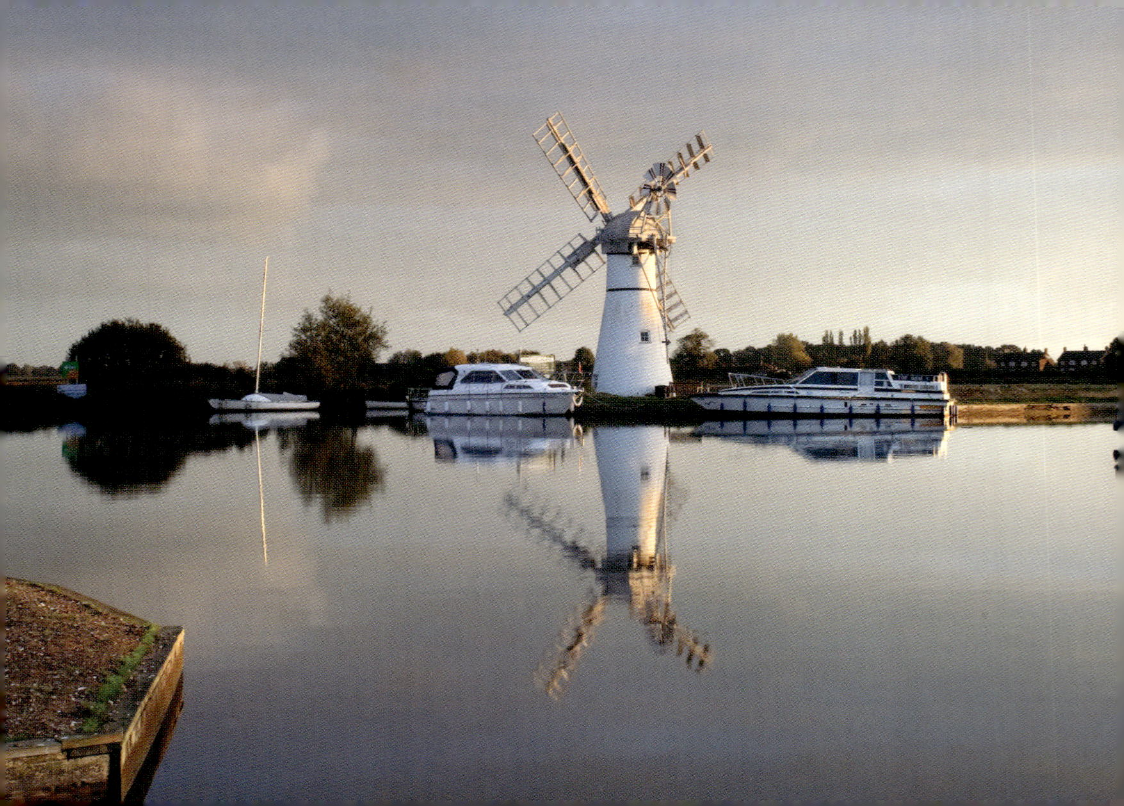

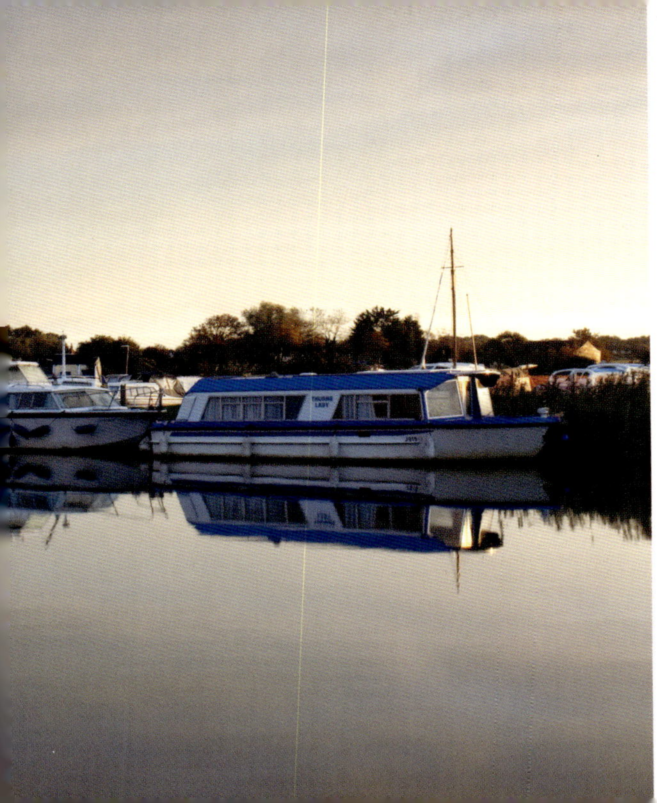

Thurne has to be my favourite of all the windmills in Norfolk and is a photographer's dream. There are so many different angles that can be exploited; here the mill and a small selection of holiday boats are captured on a calm morning reflected in the River Thurne.

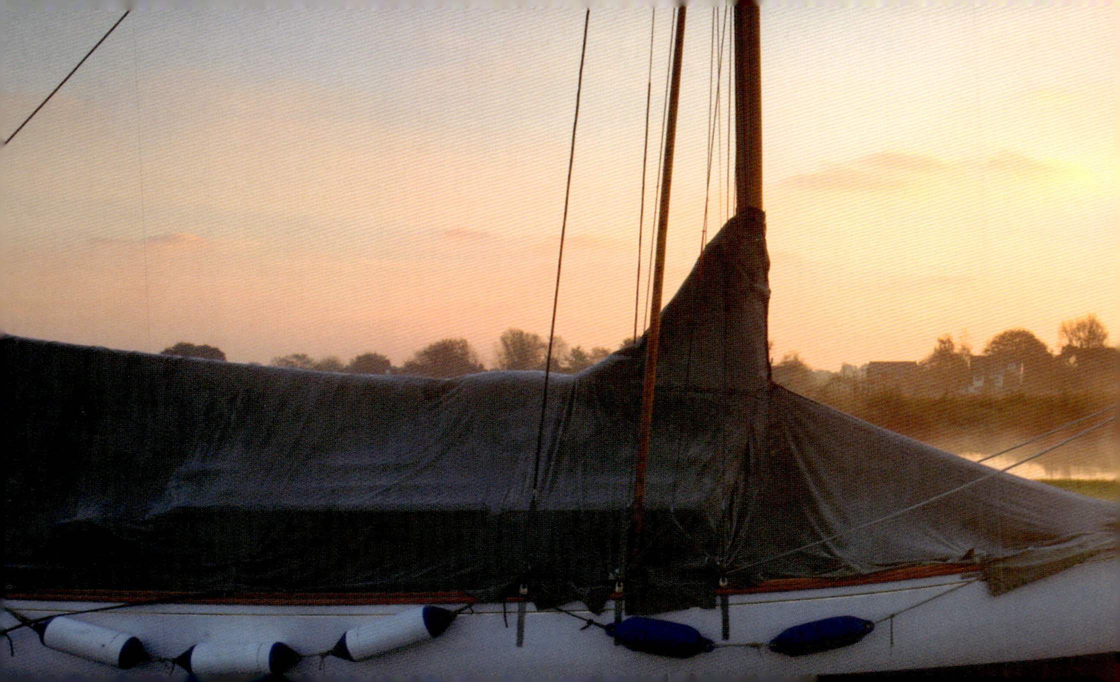

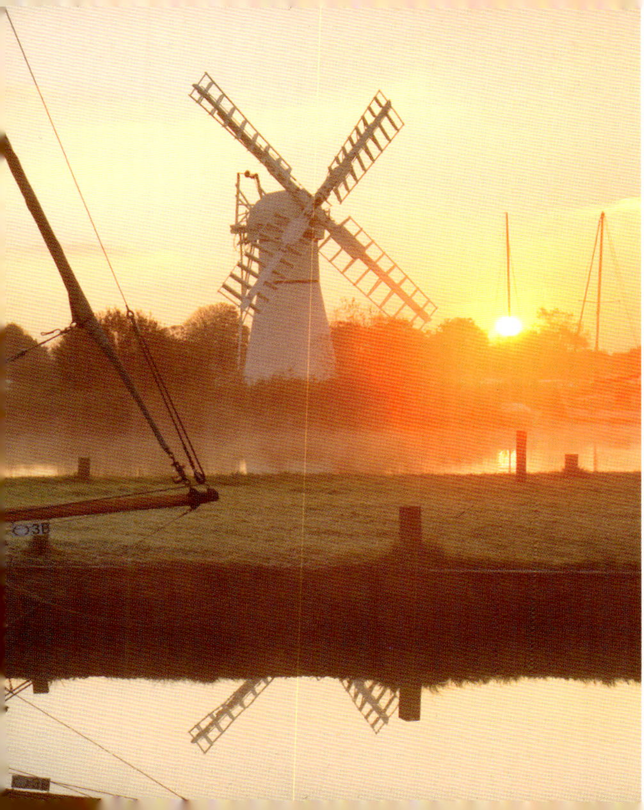

A traditional wooden sailing boat and Thurne Mill are captured here as the mist hovers above the River Thurne on an autumn morning.

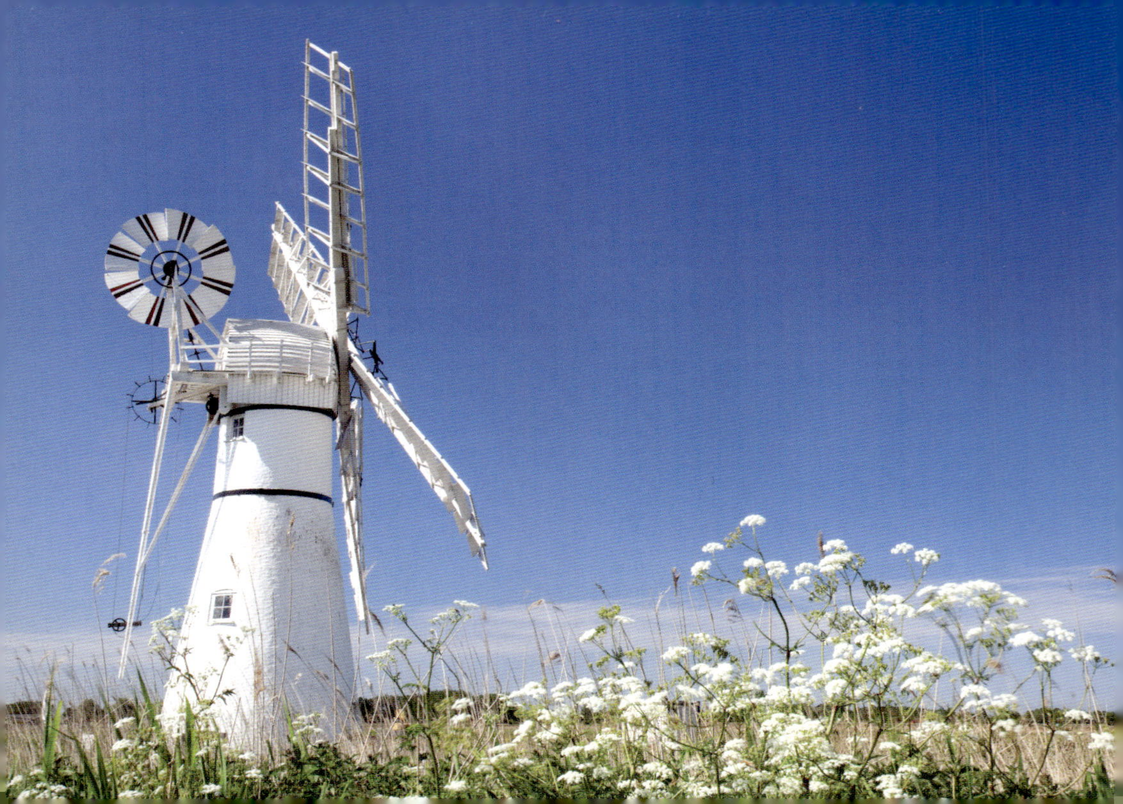

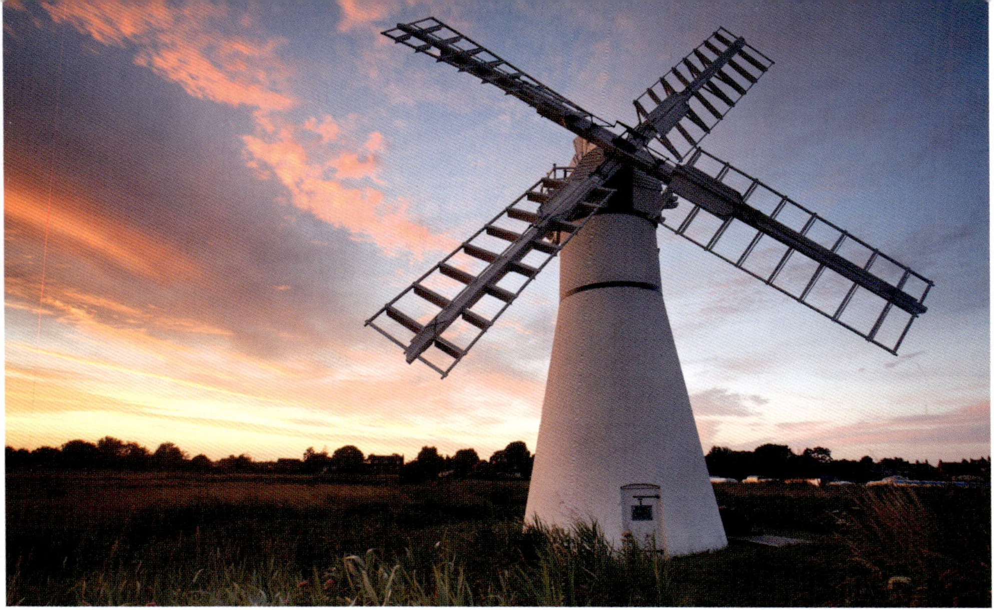

A very early start rewarded me with these spectacular dawn colours on this summer morning.

Opposite page: Thurne is one of the more popular mills with visitors on the Norfolk Broads as it is easily accessible by boat or car. Here the mill is photographed on a bright spring afternoon.

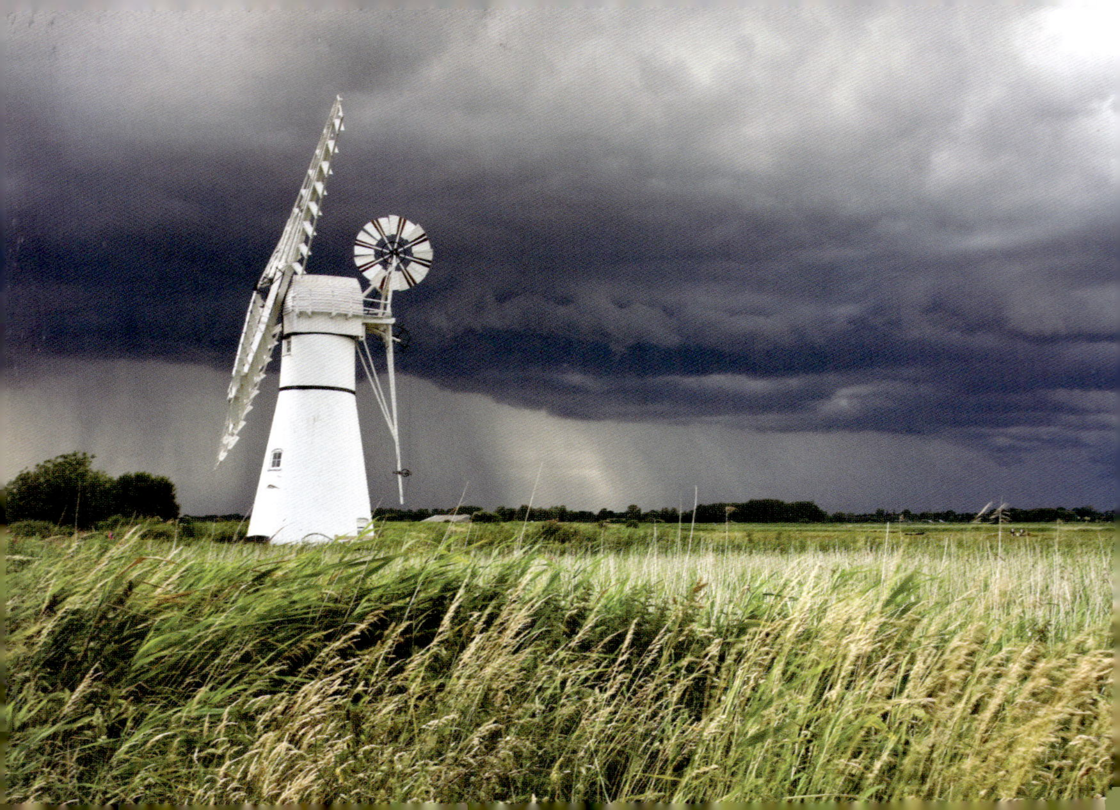

Upturned boats in front of Thurne Mill.

Opposite page:
The white-painted structure of Thurne Dyke drainage mill contrasts against a moody spring sky.

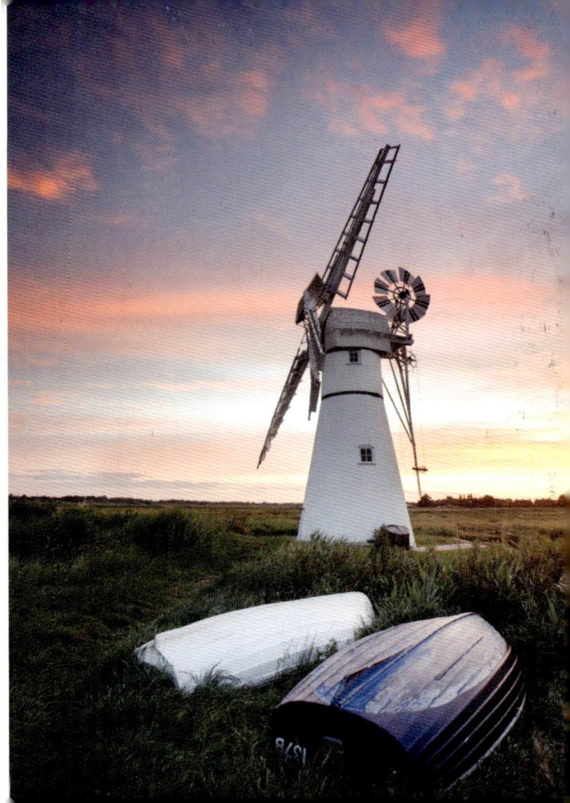

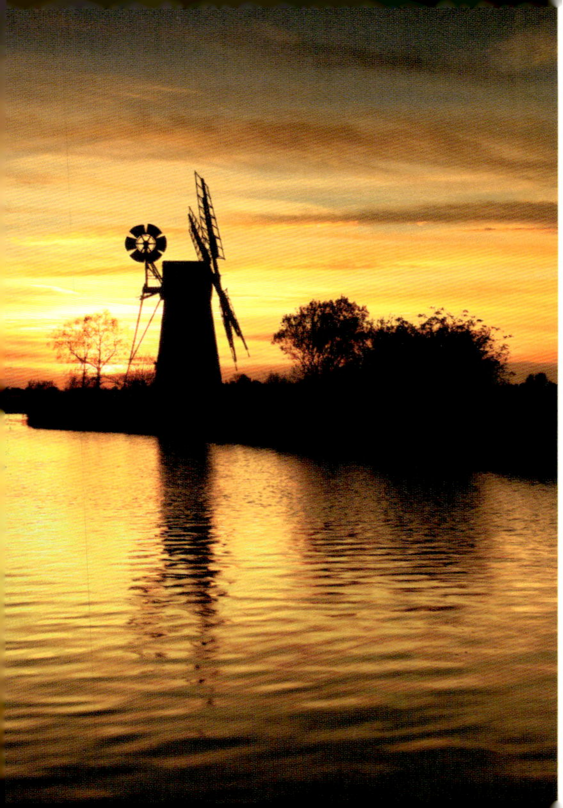

Turf Fen drainage mill at How Hill on the River Ant is captured here at sunset on a winter's afternoon.

Opposite page:
Pastel tones above Turf Fen drainage mill at dusk on the Norfolk Broads.

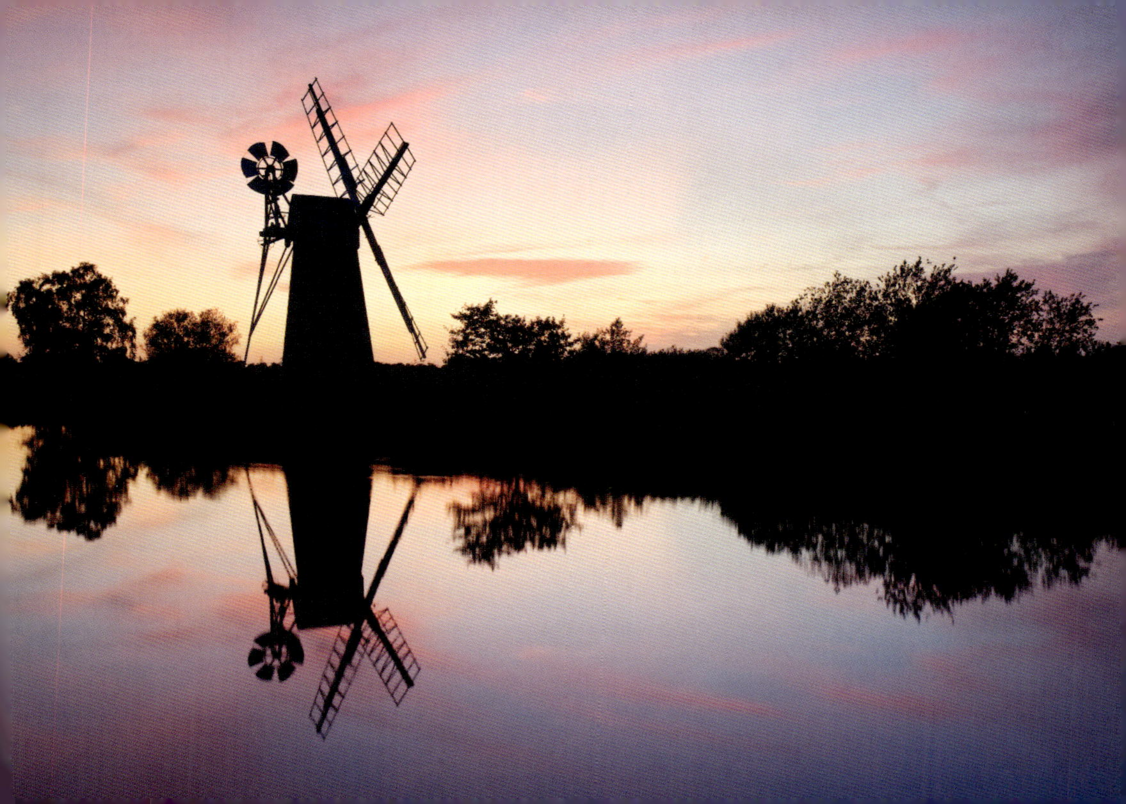

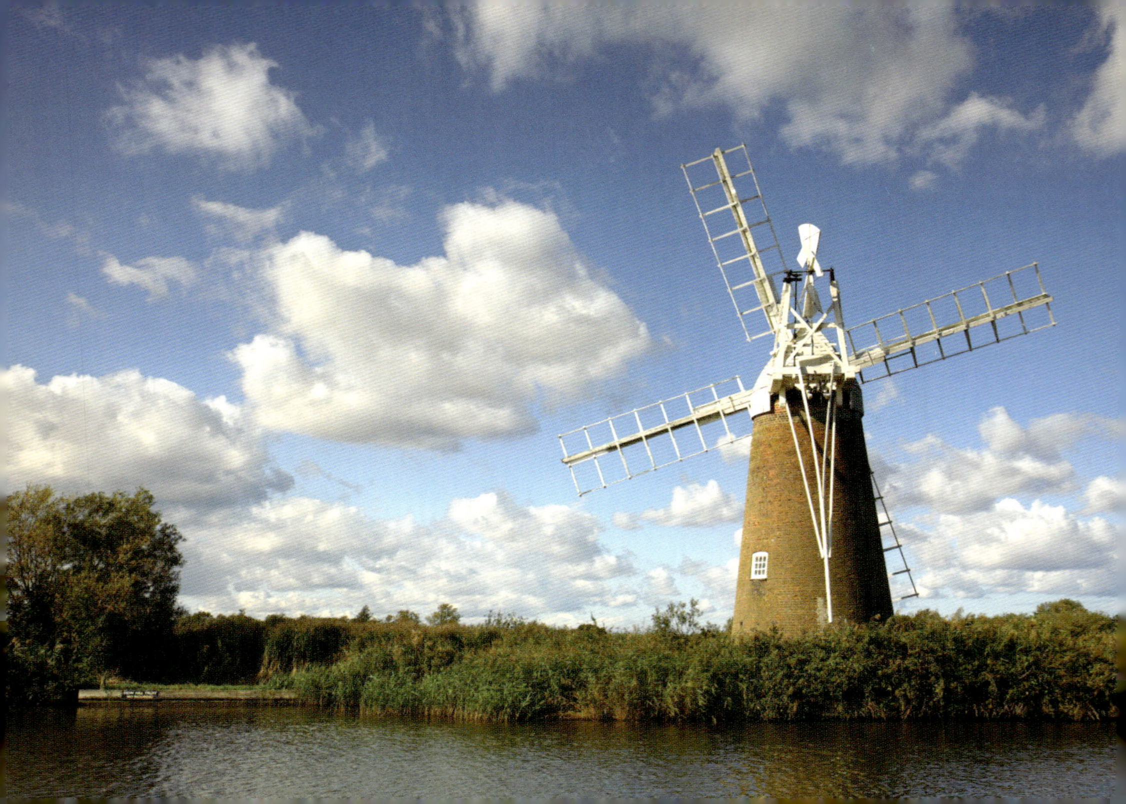

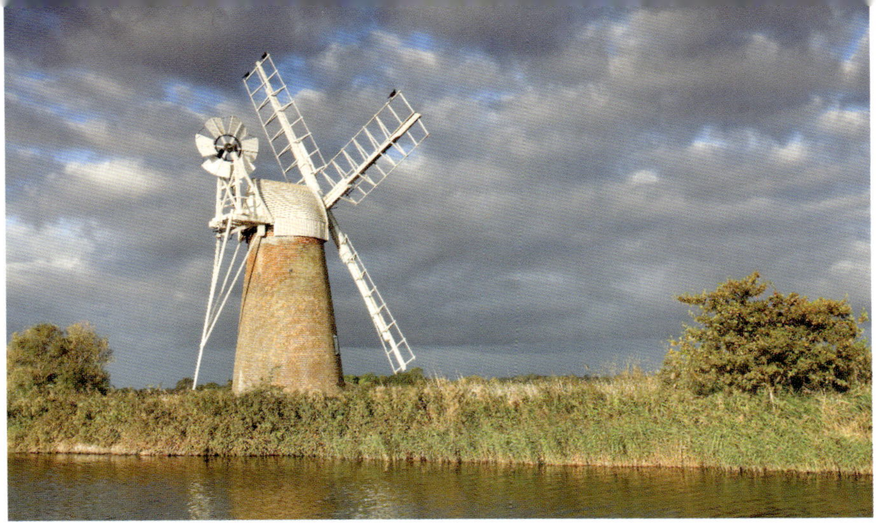

Three cormorants use the sails of Turf Fen drainage mill as a lookout point on an October morning on the River Ant.

Opposite page: Turf Fen drainage mill on the River Ant was built around 1875 by Yarmouth millwright William Rust. The mill ceased working in the 1920s when cattle no longer grazed the marshes.

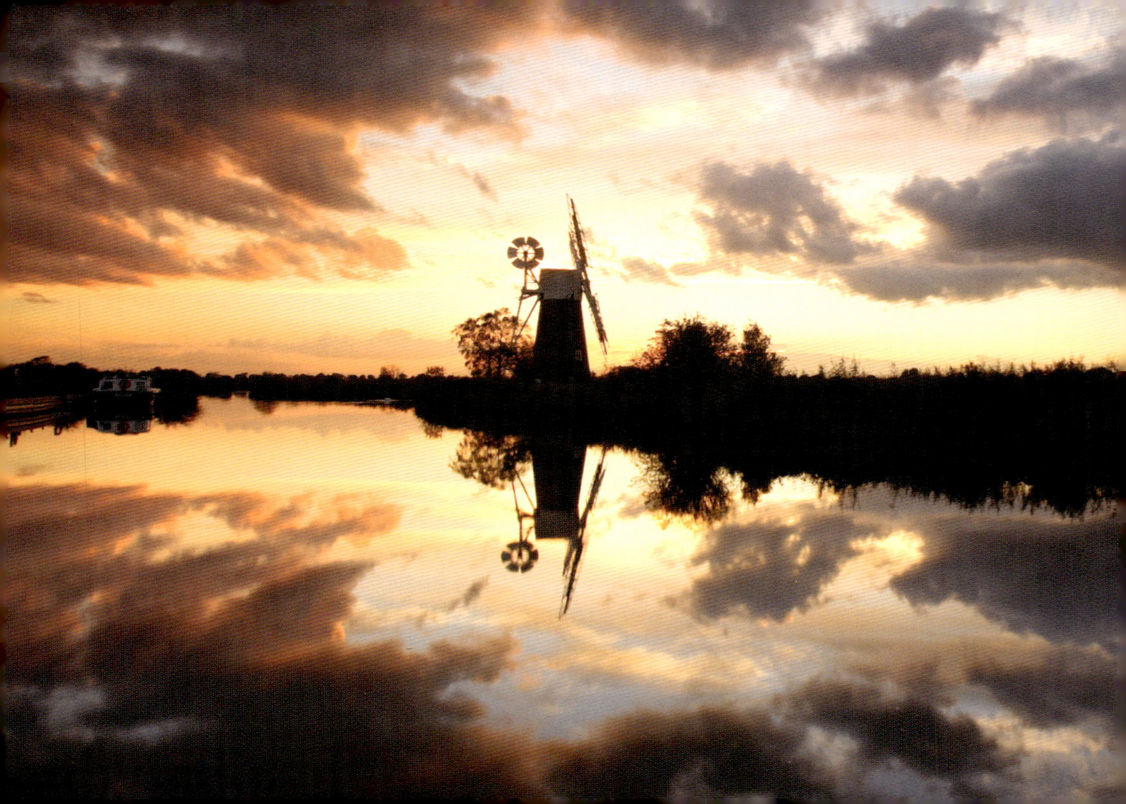

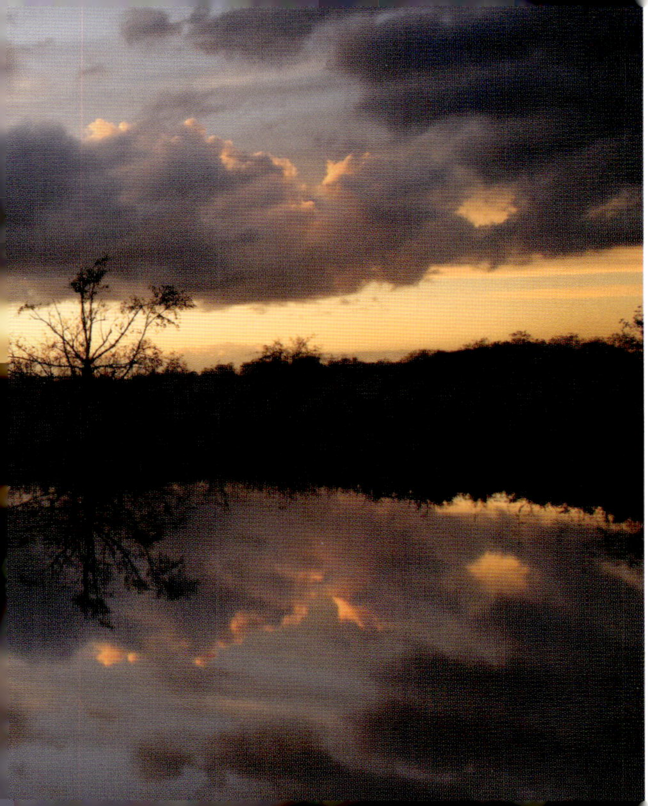

Mirror-like reflections are captured
here with Turf Fen windmill
beside the River Ant
at sunset.

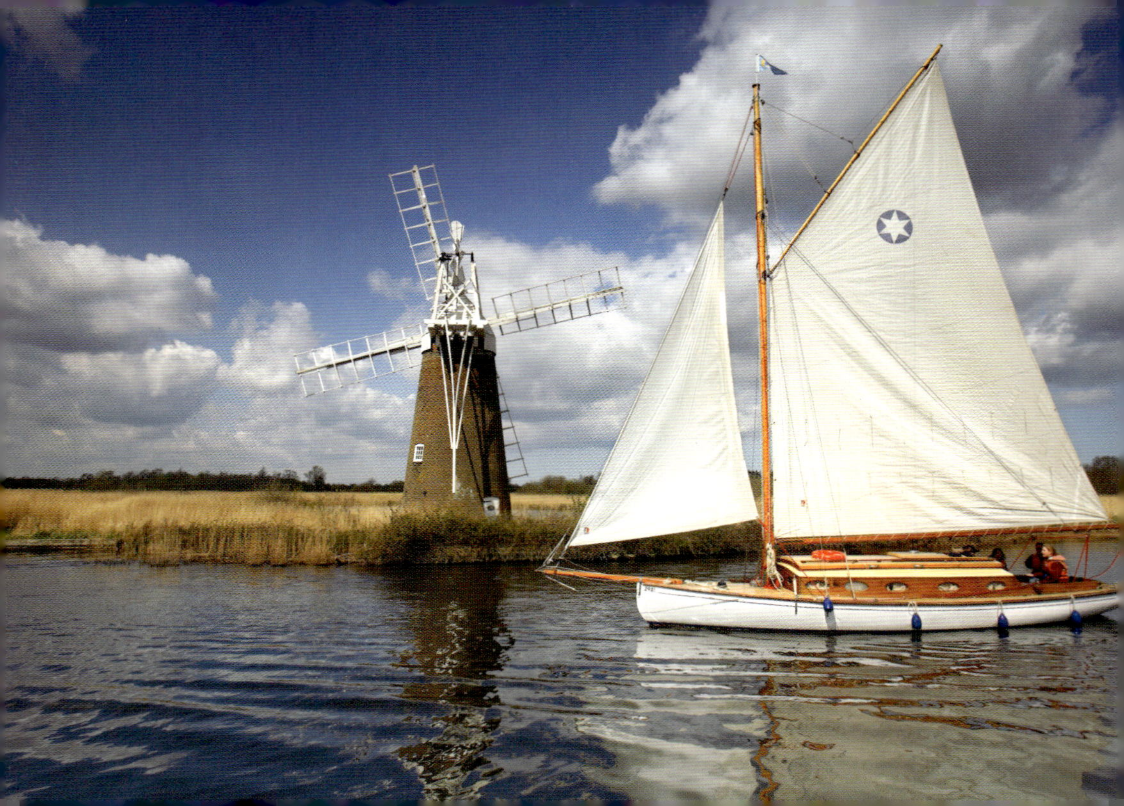

The warm winter light illuminates the winter reeds and the brickwork of Turf Fen Mill.

Opposite page:
A typical Broadland scene with a wooden sailing boat passing Turf Fen drainage mill on a spring afternoon.

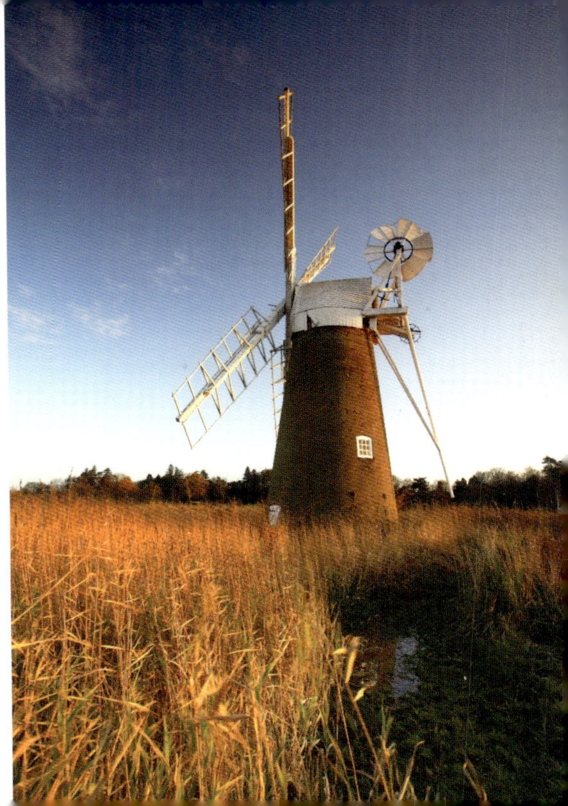

West Somerton drainage mill captured during an overhead thunderstorm on the Norfolk Broads.

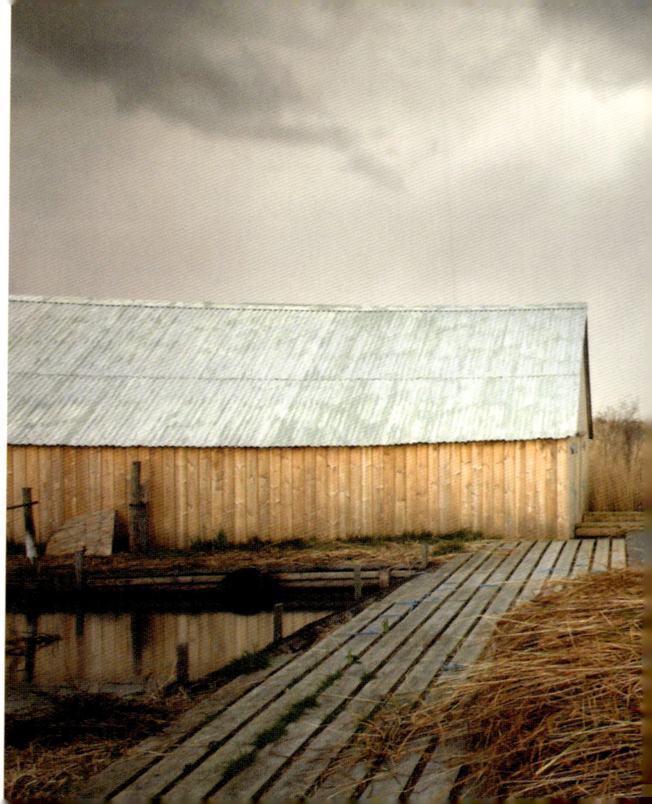

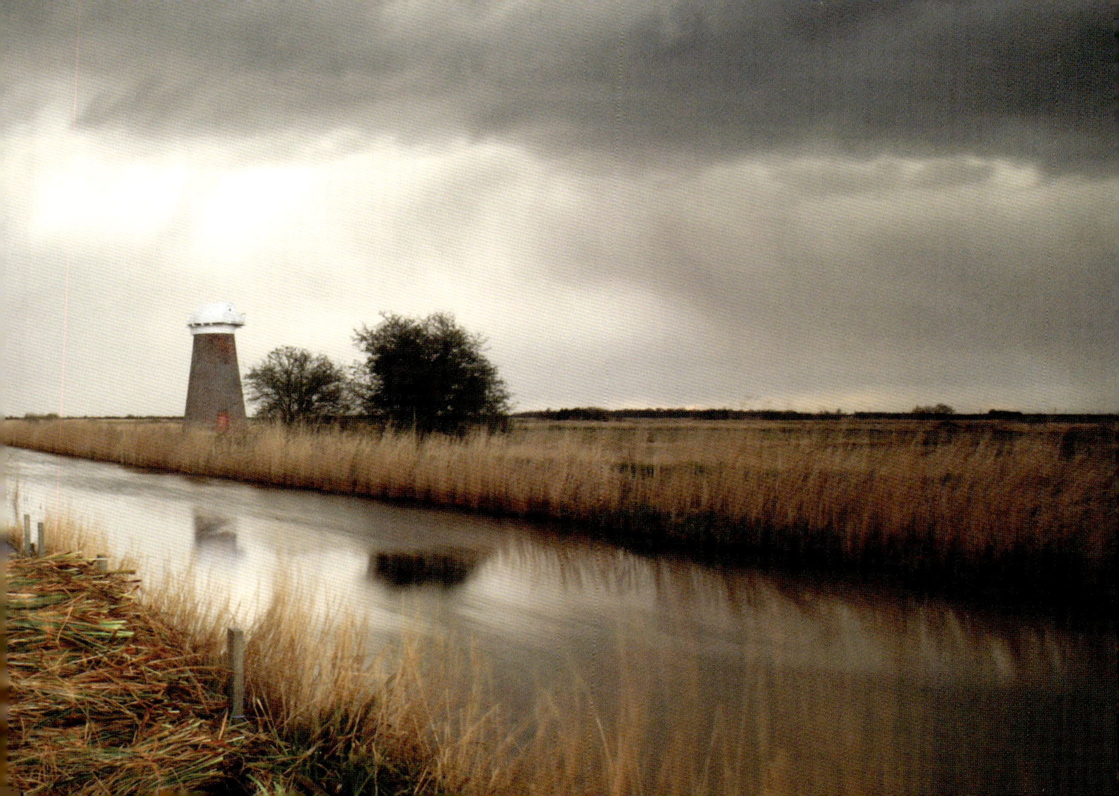

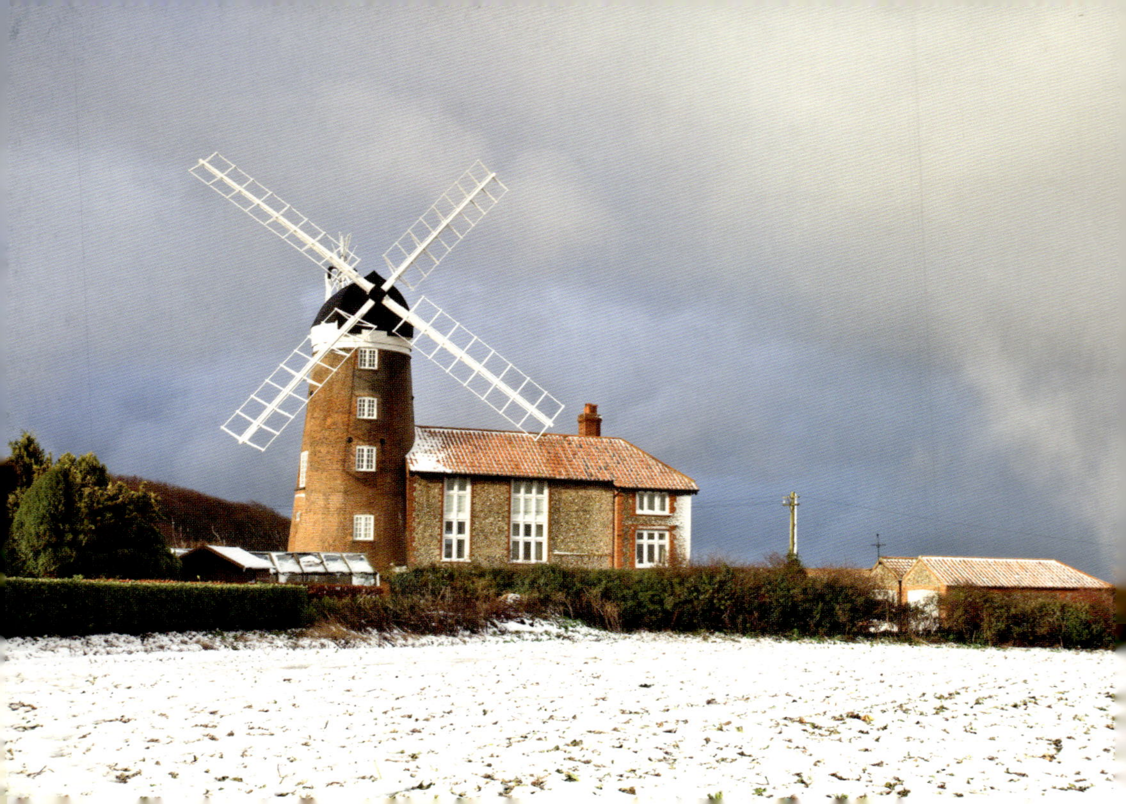

This photograph of Weybourne Mill was taken a year earlier than the image on the left and shows the main sails removed for restoration.

Opposite page:
Weybourne tower mill, seen with its newly restored sails, following a winter snow shower on the North Norfolk Coast.

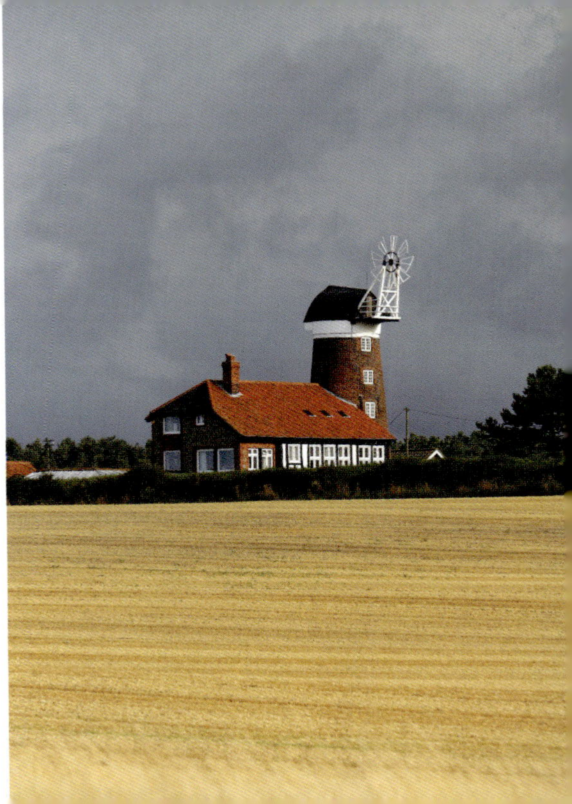

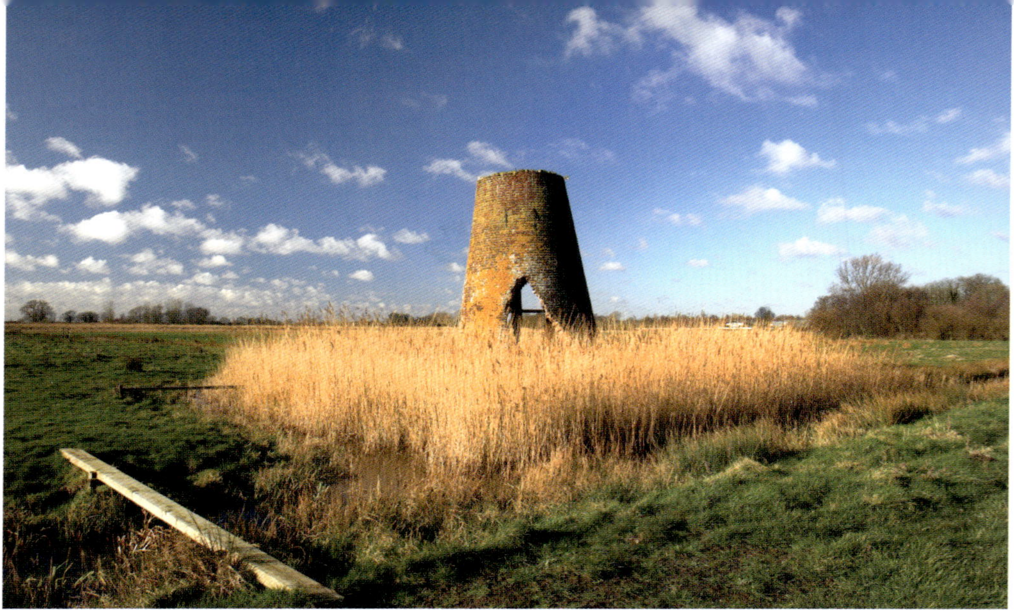

Although completely derelict there is still something very beautiful about Ludham Womack Mill. The mill is completely open with many of the original workings still intact and exposed to the elements giving a real feeling for how the mill would have once looked.